ISBN 978-1-330-42505-3
PIBN 10060910

Forgotten Books is a registered trademark of FB &c Ltd.
Copyright © 2015 FB &c Ltd.
FB &c Ltd, Dalton House, 60 Windsor Avenue, London, SW19 2RR.
Company number 08720141. Registered in England and Wales.

For support please visit www.forgottenbooks.com

1 MONTH OF
FREE
READING

at

www.ForgottenBooks.com

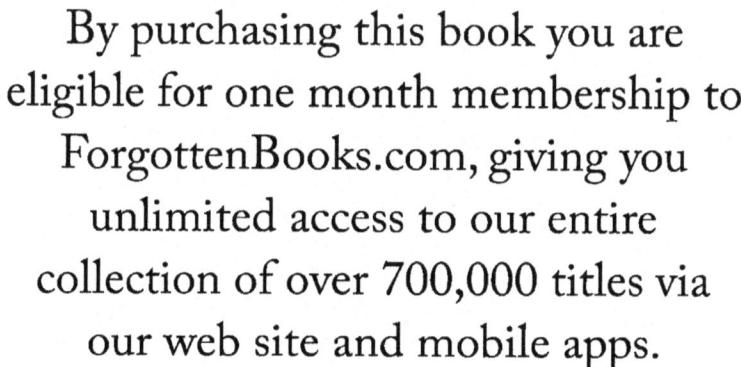

By purchasing this book you are eligible for one month membership to ForgottenBooks.com, giving you unlimited access to our entire collection of over 700,000 titles via our web site and mobile apps.

To claim your free month visit:

www.forgottenbooks.com/free60910

Similar Books Are Available from
www.forgottenbooks.com

NOTICE.

THE Committee desire to intimate that no copies of these Memoirs will be presented or exchanged, as the prices have been fixed on such a scale that most of the copies will have to be sold to meet the cost of production.

The Memoirs may be obtained, post free at the nett prices stated, from Messrs. Williams and Norgate, 14, Henrietta Street, Covent Garden, London.

Liverpool Marine Biology Committee.

L.M.B.C. MEMOIRS

ON TYPICAL BRITISH MARINE PLANTS & ANIMALS

EDITED BY W. A. HERDMAN, D.Sc., F.R.S.

VIII.)

PLEURONECTES

BY

FRANK J. COLE, JESUS COLLEGE, OXFORD,
Lecturer in the Victoria University, Demonstrator of Zoology, University College,
Liverpool ;

AND

JAMES JOHNSTONE, B.Sc., LOND.,
Fisheries Assistant, University College, Liverpool.

(With 11 Plates)

(Aided by a Grant from the Victoria University).

PRICE SEVEN SHILLINGS.

LONDON
WILLIAMS & NORGATE
DECEMBER, 1901,

EDITOR'S PREFACE.

THE Liverpool Marine Biology Committee was constituted in 1885, with the object of investigating the Fauna and Flora of the Irish Sea.

The dredging, trawling, and other collecting expeditions organised by the Committee have been carried on inter mittently since that time, and a considerable amount of material, both published and unpublished, has been accumulated. Fifteen Annual Reports of the Committee and five volumes dealing with the "Fauna and Flora" have been issued. At an early stage of the investigations it became evident that a Biological Station or Laboratory on the sea-shore nearer the usual collecting grounds than Liverpool would be a material assistance in the work. Consequently the Committee, in 1887, established the Puffin Island Biological Station on the North Coast of Anglesey, and later on, in 1892, moved to the more commodious and convenient Station at Port Erin in the centre of the rich collecting grounds of the south end of the Isle of Man. A new and larger Biological Station and Fish Hatchery, on a more convenient site is now being erected, and will, it is expected, be open for work in May, 1902.

In these fifteen years' experience of a Biological Station (five years at Puffin Island and ten at Port Erin), where College students and young amateurs form a large proportion of the workers, the want has been frequently felt of a series of detailed descriptions of the structure of certain common typical animals and plants, chosen as representatives of their groups, and dealt with by specialists. The same want has probably been felt in other similar institutions and in many College laboratories.

The objects of the Committee and of the workers at the Biological Station have hitherto been chiefly faunistic and speciographic. The work must necessarily be so at first when opening up a new district. Some of the workers have published papers on morphological points, or on embryology and observations on life-histories and habits; but the majority of the papers in the volumes on the "Fauna and Flora of Liverpool Bay" have been, as was intended from the first, occupied with the names and characteristics and distribution of the many different kinds of marine plants and animals in our district. And this faunistic work will still go on. It is far from finished, and the Committee hope in the future to add greatly to the records of the Fauna and Flora. But the papers in the present series are quite distinct from these previous publications in name, in treatment, and in purpose. They are called the "L.M.B.C. Memoirs," each treats of one type, and they are issued separately as they are ready, and will be obtainable Memoir by Memoir as they appear, or later bound up in convenient volumes. It is hoped that such a series of special studies, written by those who are thoroughly familiar with the forms of which they treat, will be found of value by students of Biology in laboratories and in Marine Stations, and will be welcomed by many others working privately at Marine Natural History.

The forms selected are, as far as possible, common L.M.B.C. (Irish Sea) animals and plants of which no adequate account already exists in the text-books. Probably most of the specialists who have taken part in the L.M.B.C. work in the past will prepare accounts of one or more representatives of their groups. The following have already promised their services, and in many cases the Memoir is already far advanced. The first Memoir appeared in October and the second in December, 1899, the third in February, and the fourth in April, 1900, the fifth in January, the sixth in March, and the seventh in April, 1901, while this eighth one will be ready in December, 1901, and the ninth in the spring of 1902; others will follow, it is hoped, in rapid succession.

Memoir I. ASCIDIA, W. A. Herdman, 60 pp., 5 Pls., 2s.

,, II. CARDIUM, J. Johnstone, 92 pp., 7 Pls., 2s. 6d.

,, III. ECHINUS, H. C. Chadwick, 36 pp., 5 Pls., 2s.

,, IV. CODIUM, R. J. H. Gibson and Helen Auld, 26 pp., 3 Pls., 1s. 6d.

,, V. ALCYONIUM, S. J. Hickson, 30 pp., 3 Pls., 1s. 6d.

,, VI. LEPEOPHTHEIRUS AND LERNÆA, Andrew Scott, 62 pp., 5 Pls., 2s.

,, VII. LINEUS, R. C. Punnett, 40 pp., 4 Pls., 2s.

,, VIII. PLAICE, F. J. Cole and J. Johnstone, 260 pp., 11 Pls., 7s.

,, IX. CHONDRUS, O. V. Darbishire.
BUGULA, Laura R. Thornely.
OYSTER, W. A. Herdman and J. T. Jenkins.
OSTRACOD (CYTHERE), Andrew Scott.
DENDRONOTUS, J. A. Clubb.
PERIDINIANS, G. Murray and F. G. Whitting.
ZOSTERA, R. J. Harvey Gibson.
HIMANTHALIA, C. E. Jones.
DIATOMS, F. E. Weiss.
FUCUS, J. B. Farmer.
BOTRYLLOIDES, W. A. Herdman.
CUTTLE-FISH (ELEDONE), W. E. Hoyle.
PATELLA, J. R. A. Davis and H. J. Fleure.
CALANUS, I. C. Thompson.
ACTINIA, J. A. Clubb.
HYDROID, E. T. Browne.

Myxine, F. J. Cole.
Calcareous Sponge, R. Hanitsch.
Arenicola, J. H. Ashworth.
Antedon, H. C. Chadwick.
Porpoise, A. M. Paterson.

In addition to these, other Memoirs will be arranged for, on suitable types, such as *Sagitta* (by Mr. Cole), a Cestode and a Turbellarian (by Mr. Shipley), *Carcinus*, an Amphipod, and a Pycnogonid (probably by Dr. A. R. Jackson).

As announced in the preface to Ascidia, a donation from Mr. F. H. Gossage of Woolton met the expense of preparing the plates in illustration of the first few Memoirs, and so enabled the Committee to commence the publication of the series sooner than would otherwise have been possible. Other donations received since from Mr. Gossage, and from Mrs. Holt, are regarded by the Committee as a welcome encouragement, and have been a great help in carrying on the work.

The greater part of the expense of the plates for the present Memoir has been defrayed by a grant from the Publications Committee of the Victoria University.

W. A. Herdman.

University College, Liverpool,
December, 1901.

L.M.B.C. MEMOIRS.

No. VIII. PLEURONECTES.

(THE PLAICE.)

BY

F. J. COLE, Jesus College, Oxford,

*Lecturer in the Victoria University, Demonstrator of Zoology,
University College, Liverpool;*

AND

J. JOHNSTONE, B.Sc., Lond.,

Fisheries Assistant, University College, Liverpool.

CONTENTS.

INTRODUCTION.

B

INTRODUCTION.

THE subject of the present Memoir belongs to the old group of fishes known as the " Teleostei," but the dispersal of the " Ganoid " fishes has necessitated a new classification, with the result that the familiar word " Teleost " may in the future be " missed from its accustomed hill " in all classifications of fishes. The Plaice (*Pleuronectes platessa*) is the most familiar example in British seas of a group having an almost universal geographical distribution. It is the typical member of this group, which has long been known as the Pleuronectidæ, a family of fishes belonging to the sub-order Anacanthini. In recent times this family has itself received sub-ordinal rank, and has been termed the Heterosomata, being divided into two families (the Pleuronectidæ and Soleidæ) and six sub-families, containing a large number of genera and species. The principal diagnostic characters of the group are the torsion which the anterior region of the skull undergoes during development, and the modification and use of the left side as the under side of the body. The lateral compression of the body is paralleled among other Teleostean fishes, but the [apparent] presence of both eyes on the right or left side of the body is a unique feature.

The nearest relatives of the Pleuronectidæ among the Teleosts are the Gadidæ, and, curiously enough, these two families afford the major portion of the fishes used as food by man. The striking differences in general body form and habits between the Plaice and Cod (typical examples of the two groups) form a marked instance of how external differences may coincide with deep seated morphological similarity. The Plaice is a fish which is sluggish in its movements, and has a very limited range of migration.

It lives, too, permanently on the sea bottom, often buried in the sand, feeds almost exclusively on bivalve molluscs, and in body form departs widely from that usual in fishes. On the other hand, the Cod is an active fish, which may migrate over wide sea areas, and although it affects the bottom, it may be found in any vertical zone of the sea. Further, it is voracious and even cannibalistic, and, although it feeds mostly on Crustacea, many marine groups of animals contribute to its food, whilst it has the typical piscine form. Nevertheless, we shall fhow that the morphological differences between the two fishes, apart from the question of symmetry, are comparatively few and unimportant.

The following parasites of the Plaice have come under our own observation :—(1) Unidentified Sporozoan cysts imbedded in the wall of the gut, and reducing it to a thin membrane ; (2) unidentified Myxosporidian cysts within the cartilage of the auditory capsule, and causing a considerable hypertrophy of the same ; (3) the Cyclops stage of *Lernœa*, attached to the gill filaments ; (4) *Chondracanthus cornutus*, inside the gill cover ; (5) *Lepeophtheirus pectoralis*, on the skin under the pectoral and pelvic fins ; and (6) *Bomolochus soleœ* in the nasal chamber. There are of course others, but these we have seen.

Only seven genera and fourteen species of Pleuronectidæ are known to inhabit British seas. The members of the genus *Pleuronectes* are *P. platessa* (plaice), *P. limanda* (dab), *P. flesus* (flounder), *P. cynoglossus* (witch), and *P. microcephalus* (lemon sole). All these species are known in the Irish Sea area. The Plaice is probably the most abundant, and the order of the species in the above list gives the relative abundance of the others. All are important edible fishes, and are the objects of an active fishing industry.

A.—EXTERNAL CHARACTERS.

The Plaice is not a large fish compared with many edible fishes, and the largest of which we find a record was 33 inches long, 21 inches high, and weighed 15lbs.† An average large plaice, however, would have an extreme measurement of 24 inches long and 14 high. The compression of the body is not from above downwards, as in the skate, but from side to side, so that when the fish is lying on the sea bottom its left side is downwards and the right only is exposed. For the sake of convenience, and for obvious reasons, we shall follow Traquair in referring to the right and left sides as the "ocular" and "eyeless" sides respectively.

The eyeless side of the Plaice is colourless and flat, whilst the ocular side is pigmented and convex. The colour varies very greatly according to the nature of the sea bottom, and may be anything from grey to dark brown. The characteristic orange red spots (ocelli) form a row of about 6 on the dorsal fin, 15 or so on the body, one on the caudal fin, and another row of about 6 on the anal fin. Specimens with the eyeless side more or less coloured, and also reversed examples, are occasionally met with. The nature of the colouration has been investigated by Pouchet,‡ and by Cunningham and MacMunn.* We follow the latter memoir. If a superficial section be made of the fresh unprepared skin of the ocular side, two structures only are apparent. These are the colour cells or chromatophores, situated largely in the dermis, but also found in the epidermis, and the opaque somewhat iridescent reflecting bodies or iridocytes. One layer of

† Thirteenth Annual Report for 1898 of Inspectors of Sea Fisheries (England and Wales), 1899, p. 10.

‡ Jour. l'anat. phys., 1877, No. 1.　　*Phil. Trans., 1893, B, p. 765.

chromatophores and iridocytes occurs outside the scales, and there is another layer of both on the inner surface of the skin and between it and the muscles. In the skin of the eyeless side no chromatophores whatever are normally present, and also only the external layer of iridocytes is found. Internally, however, there is a " thin perfectly opaque layer of material giving a dead-white reflection. Examined with the microscope, this layer presents only a minutely granular structure, and is everywhere uniform and continuous." On account of its capacity of reflecting light in such a way as to produce a silvery appearance it is called the argenteum. It is doubtful whether the chromatophores, iridocytes or argenteum are cellular structures. It may be mentioned that Cunningham and MacMunn were able by experiment to induce a colouration of the under side of the flounder.

The dorsal fin commences vertically above the left eye, a short distance behind the left posterior nostril. It extends back to the root of the tail, and is highest about two-thirds of its length from the snout. The number of fin rays varies considerably,† and in six specimens selected at random ranged from 66 to 74. The anal fin commences very far forward, immediately behind the so-called " anal spine," and stretches as far back as the dorsal fin. It is highest at about its anterior third. In the same six examples above the fin rays varied from 52 to 57. The caudal fin belongs to the masked heterocercal or homocercal type, and has usually about 20 fin rays. The pectoral fin is situated immediately below the posterior angle of the operculum. It has usually the same number of fin rays on both sides (about 10), but on the eyeless side one is small and may be overlooked. The pelvic fin is jugular

† See especially Cunningham, Jour. M. Biol. Assoc. N. S., vol. iv., 1897.

in position, slightly in front of the pectoral, and imme
diately in front of the anus. As far as we have seen its
fin rays never vary in number, and we have always found
six on both sides. This is very striking when we consider
the variability in the number of fin rays in all the other
fins.

The opercular fold or gill cover of the Plaice is large
and the branchial cavity opens behind by a wide aperture.
This aperture is bounded on the inner side by the clavicle
and on the outer side by the loose branchiostegal membrane supported by the branchiostegal rays. On lifting
the opercular fold the gill-like pseudobranch is easily
seen in a slight recess on the inner side of the operculum
immediately over the dorsal extremities of the gill arches.
Ventrally the opercular folds are separated by a conical
fleshy mass containing the " inter-clavicle," and known as
the isthmus.

The lateral line of systematists commences on the
tail, and courses straight forward at about the middle of
the side of the body for the greater part of its length. It
curves slightly upwards over the pectoral fin, and thereafter becomes buried in the bones of the head. Further
portions, however, of the lateral line system are visible on
the surface, notably the right infraorbital canal under the
right eye, and a portion of the supratemporal canal under·
the dorsal fin.

The scales are mostly cycloid, but according to
Cunningham (op. cit.) the so-called "ciliated" or
"spinulated scale"·is found only in mature males, and
may form a conspicuous local peculiarity. The scales of
this character that we have seen had three or four blunt
processes on their posterior border.

Regarding the apertures, the mouth is terminal and
markedly asymmetrical. If the mouth of a plaice be

opened, it will be seen that the whole jaw apparatus swerves towards the eyeless side. The mechanism by which this is effected is described elsewhere. The explanation of this asymmetry is that the fish must seek its food on the sea bottom, and the torsion of the jaws towards the under side is hence a physiological convenience, if not a necessity. The same consideration explains why the teeth, which are blunt and flattened, and not pointed like those of the cod, are situated almost entirely on the eyeless side. Within the mouth will be seen the maxillary and mandibular breathing valves, or "internal lips," to prevent the regurgitation of water through the mouth on the fall of the gill cover.† The anus is situated very far forward in front of the "anal spine," and is a large median opening elongated from before backwards. The urinary papilla of the female is distinctly on the ocular side, a little distance above and behind the anus. The oviducal aperture is large and lies immediately behind the anus. It is very obvious in the spawning season, but at other times of the year we have failed to find it, so that it is either occluded then, or very minute. In the immature fish we believe it does not exist. In the male the papilla is in the same position, but is here a urinogenital papilla. We have failed to find any external distinction between the two sexes, but the male is smaller than the female when it first becomes mature. The pair of anterior and posterior nostrils on each side are on the ocular side situated between and in front of the two eyes, and on the eyeless side in front and to the left of the left eye.

Between the two eyes and passing backwards there is a prominent ridge formed almost entirely by the right frontal, and in a line continued back from this ridge are

† Dahlgren, Zool. Bull., 1898.

seen five tuberosities or tubercles. These may vary considerably both in prominence and in number. Sometimes only four are present, and also any one or all may be in duplicate. As a rule, however, five occur, situated as follows:—1 and 2 on the right frontal, 3 on the right sphenotic, 4 on the right pterotic, and 5 on the right posttemporal (see figs. 1 and 21).

The Asymmetry of the Plaice.†

The most striking feature in the external appearance of the plaice, and also the most interesting in its anatomy, is the apparent presence of both eyes on the upper, right, or ocular side. But this is not the only respect in which the head of the Plaice has undergone torsion. The jaw apparatus is also very asymmetrical, and in a different direction, for whilst the eyes are twisted towards the ocular side, the jaws incline towards the eyeless side. Now it must be obvious at the outset that the asymmetry of the jaws has been superimposed on that of the eyes, and is in fact a special adaptation to an already asymmetrical fish, living on the sea bottom, and lying on its left side. We may therefore leave this asymmetry to be described in its proper place, and confine ourselves to that of the eyes.

The asymmetry of the Pleuronectidæ was first correctly explained by Traquair in 1865. The question is beset with numerous difficulties, in the form of many secondary modifications tending to mask the true course of the original torsion. Traquair, however, in his now

† We have no space to refer to the extensive literature on the asymmetry of the Pleuronectidæ, especially as the work of Traquair covers most of the facts. We should like, however, to mention an interesting paper by Holt on an abnormal sole (P.Z.S., 1894, p. 432).

classic memoir,* the facts and conclusions of which we can fully confirm, was enabled by an exhaustive examination of the skull and lateral line system to map out the exact course followed by the head before it reached its present remarkable form.

The first difficulty in the solution of the problem is the position of the anterior extremity of the dorsal fin. If this occupies the mid-dorsal line of the head, then it is obvious that the left eye must have actually passed through the substance of the head to reach the ocular side. This supposition, absurd as it may seem to us now, was in fact believed by such an observer as Steenstrup. But the anterior extremity of the dorsal fin is *not* situated in the mid-dorsal line of the head. Its skeletal support (fig. 17) and nervous supply (fig. 27.) prove conclusively, (1) that morphologically it does not belong to the head at all, and (2) that it has secondarily passed forwards over the cranium from behind. Further, an examination of the connection between the dorsal fin skeleton and the skull (fig. 17) shows us that the fin extends forwards in a straight line over the cranium without being affected in any way by the torsion of the head. (Cp. the course of the fin indicated in fig. 1.) It is therefore certain that the forward extension of the fin took place *after* the torsion was complete. Hence it does *not* occupy the median line, but follows what Traquair calls a " pseudomesial " course, and, being a purely secondary character, may be eliminated from the discussion.

The second difficulty is the mischievous assumption that the left eye has travelled over the top of the head to the right side. The *fact* is that the left eye is *not on the right side at all*. Its presence there is purely illusory. What has happened is that the *whole* of the cranium *in the region of the orbit* has rotated on its longitudinal axis to the right side, until the two eyes, instead of occupying a horizontal plane, have assumed a vertical one, and the left eye is *dorsal* to the right. *Then* the dorsal fin grew forwards over the roof of the cranium, but naturally cannot define the morphological right and left sides of the orbital region. Thus the ocular side comprises not only the right side but a portion of the left, and the true morphological median line lies *between* the two eyes and not *above* them. The relation of the eyes to the skull is, notwithstanding the rotation of the orbital region of the latter, exactly the same as in a symmetrical fish, and the only differences of importance are the atrophy of the anterior portion of the left frontal, and the purely secondary junction under the left eye of the left prefrontal and frontal (fig. 1).

Now whilst the above is a satisfactory answer to the question *how*, it does not help us with the question *why*. It is to be presumed that the first stimulus to asymmetry was an increasing tendency of the fish to lie on the sea bottom on one side of its body. Cunningham then invokes the principle of the inheritance of acquired characters, and believes that the torsion itself was produced by the action of the eye muscles. We have considered this point of view very carefully both *per se*, as a theory, and also as a supposed explanation of the facts, with the result that we cannot subscribe to Cunningham's conclusions. It is to us simply inconceivable how *any* action of the eye muscles, as they are found in fishes, could have produced the existing torsion of the head, and this quite apart from the question whether such results, if produced, would have been inherited. This, however, will be further referred to in the section on the eye. In the meantime we prefer to believe that the asymmetry of the

Pleuronectidæ has been produced by the action of natural selection, *i.e.*, by the accumulated effects of congenital variations.

B.—THE SKELETON.

This may be divided, as usual, into an axial and an appendicular skeleton. We shall describe the former first, but the precise order must to a certain extent depend on convenience rather than upon strict logical sequence. We therefore begin with the cranium itself, afterwards proceeding to the remaining constituents of the skull, then to the vertebral column and unpaired fins, and finally to the limb girdles and paired fins.

1.—Cranium† (Figs. 1 to 4).

Owing to the difficulty of making an independent preparation of the chondrocranium it is, in the following description, described in the pieces into which it is divided when the cranium is disarticulated.

Seen from behind (Pl. II., fig. 4) the occiput is markedly asymmetrical, and a line traversing median structures would be convex towards the ocular side. This is observable also in the occipital condyle and in the paroccipital condyles (*O.C.*, *P.C.*), and of the latter, the eyeless one, as may be assumed from the description of the atlas, is larger than the ocular. In those two extensions of the auditory capsule, the epiotic and parotic processes

---

† Cp. especially, *Traquair*, Trans. Linn. Soc., xxv., p. 263. Space forbids a discussion of the literature in the text, but the following papers should be consulted :—*Schmid-Monnard*, Jena. Zeits., xxxix. ; *T. J. Parker*, Trans. Zool. Soc., xii., p. 5 ; *Sagemehl*, Morph. Jahrb., ix., p. 177 ; x., p. 1 ; xvii., p. 489 ; *Shufeldt*, Report U. S. F. C., 1883, p. 747 ; *Allis*, Jour. Morph., xii., p. 487 ; xiv., p. 425 ; Zool. Bull., i., p. 1 ; Anat. Anz.,xvi., p. 49 ; xvii., p. 433 ; *Cole*, Trans. Linn. Soc., ser. ii. vii., p. 115 ; *W. K. Parker*, Phil. Trans., vol. 173, pp. 139 and 443 : vol. 163, p. 95 ; *McMurrich*, Proc. Canadian Inst., N. S., ii., p. 270 ; *Brooks*, Sci. Proc. R. Soc., Dublin, N.S., iv., p. 166.

(fig. 4, *Ep.P.*, *Pa.P.*), there is a marked difference from the cod, the former being more prominent than the latter, instead of vice versâ.

In a dorsal view (fig. 1) the asymmetry is most pronounced in front of the parietal region. In the cod the frontals completely meet (and indeed fuse) in the mid-dorsal line. In the Plaice, on the other hand, there is a very wide separation of the frontals anteriorly, so as to form a large *secondary* frontal fontanelle or left orbit.

The asymmetry is, however, more evident on the ventral surface (fig. 2). This is due to the fact that in front of the alisphenoids there is no side wall to the cranium, which, therefore, here consists of the parasphenoid only. In front of the prootic the parasphenoid turns sharply towards the eyeless side to such an extent that the head of the vomer was, in the specimen figured, deflected by about a centimetre from the middle line. As the parasphenoid is the most prominent feature on the base of the cranium, the appearance of torsion in this region is, in a full-sized fish, most striking.

Seen from the side the cranium on the eyeless side falls more into one plane than on the ocular, but this is obviously due to the inclination of the parasphenoid and vomer to that side (cp. fig. 2).

The interior of the brain case is extremely irregular. Owing to the lateral walls meeting ventrally at a some what acute angle a false floor for the brain becomes neces sary, and this consists of two distinct parts. In front there is a rather narrow transverse bridge connecting the two alisphenoids, a strong sutural union being effected in the middle line. Behind there is a similar but much broader bridge joining the two prootics, the two processes meeting as before in a median suture. The true floor of the cranium is formed in the former of these cases by the parasphenoid and in the latter partly by the prootic and partly by the parasphenoid. In both cases there is an obvious space between the false and true floors, and this space is the eye muscle canal. At the region of this posterior bridge the side walls of the cranium are greatly strengthened internally (and the cranial cavity hence reduced) by a stout ridge of bone borne on the prootic, sphenotic and supraoccipital. From the middle of this ridge there extends backwards another process which becomes larger and more complex as it passes backwards. This is formed mostly by the sphenotic, supraoccipital, pterotic, epiotic and exoccipital, and consists of both bone and cartilage. It is here that the cranial wall is thickest. The foramen magnum does not open at once into the cranial cavity, but into a bony canal formed by the basioccipital and exoccipitals (fig. 4).

Basioccipital (*B.O.*, figs. 2, 3, 4).—A stout bone, partly cartilaginous in front, and bearing the single concave occipital condyle for the centrum of the atlas. Above it forms a small portion of the floor of the foramen magnum. Mid-ventrally it exhibits a deep depression into which fits the posterior extremity of the parasphenoid. The basioccipital is bounded above by the exoccipitals, and laterally by the prootics, opisthotics and exoccipitals.

Exoccipital (*Ex.O.*, figs. 2, 3, 4).—Forms most of the occipital foramen or foramen magnum (fig. 4, *F.M.*). It is not completely ossified, and above its cartilage forms part of the cross-shaped wedge of cartilage appearing on the surface of the occiput (fig. 4). Each exoccipital bears a very prominent ridge and concave facet lined with cartilage for the corresponding process on the atlas. The asymmetry of these paroccipital condyles (*P.C.*) has been elsewhere noticed. The exoccipital is bounded above by the epiotic, laterally by the pterotic and opisthotic, below by the basi-occipital and opisthotic, and internally by its fellow of the opposite side. It bears a conspicuous foramen with a long canal (*f.vg.*, figs. 2, 3) for the exit of the vagus nerve, and at least one other for the first spinal nerve.

Supraoccipital (*S.O.*, figs. 1, 4).—A large asymmetrical bone hardly appearing on the occiput, from which it is excluded by the epiotics, and having only a feeble occipital spine (*Oc.S.*), so well marked in the Cod. This spine and ridge is continued forwards to the left anterior corner of the bone, where it forms a furrow developed in connection with the interspinous bones or axonosts of that part of the dorsal fin situated over the head, as elsewhere described. In front the supraoccipital is thin and laminate, so that the roof of the cranial cavity is here very slender, but behind the cerebral surface of the bone is supported by three strong ridges of bone and cartilage. The supraoccipital is bounded in front by the frontals, laterally by the parietals, and behind by the epiotics. The basi-, ex- and supraoccipitals together form the occipital segment of the cranium.

The **Auditory Capsule** of the Plaice is formed by the following five bones, as in the Cod:—

Sphenotic (*Sp.O.*, figs. 1, 2, 3).—Does not contain much cartilage. Externally on the dorsal surface a strong process is sent out and supported by a ridge of bone coming up from below. The sphenotic forms the external and upper half of the deep cup for the ball of the hyomandibular (cp. fig. 5), the prootic half of the same being more or less separated from it by a strip of the chondrocranium (cp. the two sides in fig. 2, *Hm.F.*[1]). The cerebral surface of the sphenotic has two large cavities separated by a thick wall of bone and cartilage. The sphenotics are not quite symmetrical—that of the ocular side being the larger and more densely calcified. They are bounded by the frontal, alisphenoid, prootic, pterotic and parietal.

Prootic or **Petrosal** (*Pr.O.*, figs. 2, 3).—A stout bone containing a quantity of cartilage. It is perforated by the large canal or foramen jugulare (*f.jug.*), which transmits the internal jugular vein, the ophthalmic artery and the truncus hyomandibularis nervi facialis. It also forms the postero-lateral wall of the trigemino-facial foramen (*f.tr.fa*) and the external wall of the carotid foramen (*f.car.*), transmitting the internal carotid artery. Further it forms the internal and lower half of the hyomandibular cup (*Hm.F.*[1]), and its part in forming a false floor to the cranial cavity by processes of bone and cartilage has been already mentioned. The prootic has two conical depressions on its cerebral surface, the ventral one being much the larger. It is bounded by the parasphenoid, alisphenoid, sphenotic, pterotic and basioccipital.

Epiotic (*Ep.O.*, figs. 1, 3, 4).—A dense structure largely cartilaginous, but having a thin outer shell of bone, prolonged into the somewhat prominent epiotic process (*Ep.P.*, fig. 4). The cerebral surface bears two or three deep conical pits with a thin bony lining, one being much larger than the others. The epiotic provides the remainder of the cartilage for the occipital cross already mentioned. It is bounded by the supraoccipital, parietal, pterotic and exoccipital.

Pterotic (*Pt.O.*, figs. 1, 2, 3).—Forms the greater part of the parotic process (*Pa.P.*, fig. 4). It is more densely calcified than the epiotic, and its cerebral surface bears three deep conical pits, one being partially subdivided into two. Laterally it bears an imperfect oval bony facet for the posterior condyle of the hyomandibular (*Hm.F.*[2], figs. 2, 3, and cp. fig. 5). The left facet is appreciably larger than the right (cp. description of the hyomandibu-lar), in contradistinction to the hyomandibular cup which is smaller on this side. The pterotic forms the anterior boundary of the glosso-pharyngeal foramen as shown in fig. 2 (*f.gl.*). It is bounded by the parietal, sphenotic, prootic, opisthotic and exoccipital.

Opisthotic or **Intercalar** (*Op.O.*, figs. 2, 3, 4).—Forms the remainder of the parotic process. It is by far the smallest of the otic bones, consists of a thin flat plate of irregular shape, and contains no cartilage whatever. Its development should therefore be studied. It forms the posterior boundary of the glosso-pharyngeal foramen (*f.gl.*, figs. 2, 3), and is bounded by the basioccipital, pterotic and exoccipital.

There can be no question in forms such as the Cod and Plaice that the ear bones described as pterotic and sphenotic are something more than what they seem, *i.e.*, they have a compound and not a simple origin. Added to the so-called cartilage bone in each case is a dermal element, originally developed around that part of the sensory canal system associated with these bones, and now more or less completely fused on to them. In some Fishes (such as the sphenotic of *Amia*) the two portions remain distinct throughout life, and in others the line of fusion may be plainly seen, with, however, the bones remaining separate as an occasional abnormality (such as the pterotic of the Cod). But as a rule the two portions fuse completely, so as to be indistinguishable in the adult. Now in the one case the terms pterotic and squamosal have been applied indifferently to the compound of the adult. We may therefore, in those cases where the two parts of the compound remain separate in the adult, call the cartilage pterotic, or ear bone, the true pterotic, and the dermal pterotic, or lateral line bone, the squamosal. In the other case, however, in which the terms sphenotic and post-frontal are synonyms, we cannot adopt the same plan, since the term post-frontal cannot be correctly applied to a membrane bone in Fishes. We must hence distinguish between the cartilage true sphenotic, or ear bone, and the dermal sphenotic, or lateral line bone, without giving the latter a definite name. The subject would repay investigation.

Parietal (*Par.*, figs. 1, 3).—Flat conspicuous bones containing of course no cartilage. On the dorsal surface the inner portion is laminate, but the outer portion is much more densely calcified (cp. fig. 1). The boundary separating these two parts is where the skull begins to shelve down. The two parietals are markedly asymmetrical, as shown in fig. 1. The parietal is bounded by the supraoccipital, frontal, sphenotic, pterotic and epiotic.

Alisphenoid (*Al.S.*, figs. 2, 3).—Forms, as described above, a false floor to the cranial cavity, separating the latter from the eye muscle canal. The greater part of the dorsal portion of the alisphenoid consists of two thin plates of bone with a layer of cartilage between them. Behind, the alisphenoid forms the anterior boundary of the foramen for the fifth and seventh cranial nerves, and it is at this region that the bone is most densely calcified. It is bounded by the parasphenoid, prootic, sphenotic, and frontal. In front a portion of the border is free.

Anterior to the parietal region the asymmetry of the skull is most emphasized, and its rotation in the direction of the hands of a watch is quite manifest. The bones of the two sides therefore differ more or less considerably.

Right Frontal (*R.Fr.*, figs. 1, 2, 3).—Very elongated from before backwards and narrowed from side to side. It is the anterior prolongation of the right frontal that forms the stout bar between the eyes so prominent in the

c

undissected fish. The right frontal, like the left, is
bounded by the frontal of the other side, supraoccipital,
parietal, sphenotic, alisphenoid the prefrontal of its side
and median ethmoid, except that the left frontal does not
reach the median ethmoid.

Left Frontal (*L.Fr.*, figs. 1, 2, 3).—Takes no part in
forming the interorbital ridge. Compared with the right
frontal it is broad from side to side and shorter from
before backwards. As shown in figs. 1 and 2, it sends out
on the right a strong transverse process which overlaps
the dorsal surface of the right frontal. The forward
process on the left of the left frontal lying over the left
prefrontal, together with the posterior portion of the latter
bone, are lying apparently in a very anomalous position,
i.e., they are situated under the left eye instead of over it.
This, however, has been produced by the frontal growing
forwards, and the prefrontal growing backwards, *after* the
torsion of the cranium was an accomplished fact. It is
thus a precisely analogous case to the anomalous position
of the anterior extremity of the dorsal fin.

Prefrontal (*R.* and *L. P.Fr.*, figs. 1, 2, 3).—The left
prefrontal is in every respect larger than the right—due
apparently to the circumstance just mentioned. Both
contain in front some cartilage which is continuous with
what we have termed the ethmoid cartilage. Both also
are perforated by a foramen transmitting the olfactory
nerve to the olfactory laminæ of the nasal organ, the left
foramen being perceptibly smaller than the right—due
to the left olfactory nerve being so much smaller than the
right. The left prefrontal fits by means of a long narrow
backward process into a groove on the dorsal surface of the
front end of the parasphenoid—a process entirely absent
on the right side. The articular surface for the lachrymal
is smaller on the left side, and similarly the process

bearing it is also the smaller of the two (cp. fig. 1). Above the olfactory foramen the left prefrontal is prolonged upwards and backwards to assist the median ethmoid in forming the anterior boundary of the left orbit. This process is practically absent on the right side. The prefrontal, which is called bv other authors ectethmoid, lateral ethmoid, parethmoid, or paired ethmoid, is bounded by the lachrymal, vomer, mesethmoid, ethmoid cartilage, and frontal, the left one further by the parasphenoid.

Mesethmoid (*M.E.*, *M.E.*[1], figs. 1, 3).—Presumably an ossification of the ethmoid cartilage. In front it bears a prominent beak (*M.E.*[1]), and above this is a depression, both of which are connected with the motion of the inter-maxillary cartilage. The marked inclination of the former to the eyeless side will be noted, thus diverting the motion of the jaws to that side. Behind, the mesethmoid is laminate, and takes a sharp turn upwards, forming by a graceful curve the anterior wall of the left orbit, the remainder of which is contributed by the left prefrontal. In front and on each side of the mesethmoid the ethmoid cartilage appears on the surface of the cranium, whilst behind and above on the right is an attachment for the right nasal. The mesethmoid is bounded by the vomer, ethmoid cartilage, prefrontals, right nasal and right frontal.

Ethmoid Cartilage (*Eth.*, figs. 1, 2, 3).—An asymmetrical unpaired cartilage quite separable in the macerated skull except for those portions embedded in the prefrontals. It consists of two parts, one a long basal horizontal rod tapering to a point behind and fitting into a deep groove on the upper surface of the parasphenoid alongside the caudal process of the left prefrontal, the other a vertical plate with a marked deflection to the ocular side and perforated by a large fenestra. The latter transmits the origins of the two right oblique muscles of the eye. The ethmoid cartilage appears on the surface of the skull on each side of the mesethmoid immediately above the vomer. It is bounded by the parasphenoid, the prefrontals, mesethmoid and vomer.

Vomer (*Vo.*, figs. 1, 2, 3).—A median unpaired bone consisting of an anterior head and a posterior shaft tapering to a point. The latter is firmly fixed into a long tapering cavity in the base of the parasphenoid to such an extent that the extremity of the parasphenoid is brought very near the anterior end of the vomer. The cavity in the parasphenoid lodging the vomer is quite distinct from that immediately above it for the ethmoid cartilage and the left prefrontal. The head of the vomer is markedly asymmetrical, and has a laminate process on each side, the right of which is appreciably larger than the left. In front the vertical face is inclined towards the eyeless side, thus further deflecting the motion of the intermaxillary cartilage, and hence the jaw apparatus, to that side. The vomer is bounded by the parasphenoid, prefrontals, ethmoid cartilage and mesethmoid.

Parasphenoid (*Pa.S.*, figs. 1, 2, 3).—A very long unpaired bone with a prominent keel. It is very asymmetrical, taking at the region of the alisphenoids a sharp turn towards the eyeless side. Behind it fits into a depression on the base of the basioccipital, and forms a portion of the floor of the cranial cavity in front of the latter bone, its dorsal surface being here deeply grooved. Its relations in front to the ethmoid cartilage, left prefrontal and vomer have been described above. The parasphenoid is bounded by the basioccipital, prootics, alisphenoids, left prefrontal, ethmoid cartilage and vomer.

Nasal (*R.Na.*, figs. 1, 2, 3).—Only the right nasal is present—the left having completely aborted with the almost complete loss of the left supraorbital sensory canal, the anterior extremity of which it is its main function to protect.† The existence of the left nasal would of course also be jeopardised by the motion of the intermaxillary cartilage over the beak of the mesethmoid, which, whilst not affecting the right nasal, would tend to reduce the left. It is necessary to assume some co-operative cause such as this, since the disappearance of a sensory canal does not necessarily involve the reduction of the true lateral line bones, or the Plaice would not possess a right lachrymal. The nasal of the Plaice is a small semilunar bone attached to the right side of the posterior vertical plate of the mesethmoid. It supports the anterior extremity of the right and only supraorbital sensory canal, and bounds the right nasal sack internally. It is sometimes called the turbinal.

Lachrymal (*R.Lc.*, *L.Lc.*, figs. 1, 2, 3).—These have been modified from the first of the suborbital series or chain of lateral line ossicles supporting the infraorbital sensory canal, and may hence be called the first suborbitals. They have also been called the adnasal bones, on account of their relations to the nasal sack. In the Plaice they differ from the bones of the same name in most Teleosts (including the Cod) in being closely attached to the cranium. They differ in shape as shown in fig. 1, the left being more concentrated than the right (the latter best shown in fig. 3). The right lachrymal has no connection whatever with the right infraorbital sensory canal. Both lachrymals bound their nasal sack externally,

---

† Traquair (loc. cit., p. 284) describes a "minute turbinal [nasal] ossicle" on the left side, supporting the "remnant of the main [supra-orbital] canal of the eyeless side." We have found the latter in our sections as Traquair describes it (see elsewhere), but not the rudimentary nasal. Dr. Traquair's work, however, is so accurate, that he is doubtless correct in this also.

and are both attached to anterior projections from the prefrontals in the manner described above.

The suborbital and supratemporal chains of lateral line ossicles are described with their respective sensory canals.

2.—The Palato-Pterygoid Arcade (Fig. 5).

Ocular Side.

Hyomandibular (*Hm.*).—This bone articulates with the skull in two ways. First bv a well marked ball and socket joint situated at the anterior extremity of its articular surface, and second by a long and less conspicuous facet behind this. The socket is a deep depression very obvious in dried skulls and situated between the sphenotic and prootic. From this depression the other facet passes upwards and backwards, and is placed mostly if not entirely on the pterotic. The head of the hyomandibular is cartilaginous in three places, at the two facets for the skull and at the projection articulating with the operculum. Parallel with the posterior edge, and at a short distance from it, is a stout bony ridge (the most strongly calcified part of the bone) which is closely attached to the pre-operculum. In front of this a shaft of cartilage passes downwards and forwards, which bears a thin cartilaginous cap, and articulates with the interhyal in such a way as to admirably illustrate what is now a commonplace of vertebrate morphology, that the hyomandibular is the modified dorsal segment of the hyoid arch, which has in many forms lost its connection with the hyoid arch, and has acquired on the one hand a connection with the auditory capsule and on the other with the jaw suspensorial apparatus. The Plaice is therefore hyostylic. The remainder of the ventral·edge of the hyomandibular articulates with the meta-pterygoid. In front of the cartilaginous shaft it simply consists of a thin leafy plate—a part which is absent in the Sole according to Cunningham.

Symplectic (*Sy.*).—Consists of a cartilaginous shaft with a semilunar plate of bone opposed to its anterior edge. Its head forms with the shaft of the hyomandibular the cup for the upper end of the inter-hyal, and also bears a cartilaginous epiphysis. Ventrally its extremity lies under the quadrate. Its anterior bony margin is attached to the metapterygoid and its upper posterior edge to the pre-operculum.

Quadrate (*Qu.*).—A laminate bone bearing posteriorly a strongly calcified ridge and ledge for the pre-operculum. Dorsally it is prolonged into a spine situated in front of the pre-operculum and wedged in between that bone and the symplectic. In front it articulates with the pterygoid, and below its free extremity bears a stout knob with a saddle-shaped articulation for the articular.

Meta-pterygoid (*M.Pt.*).—A very thin leafy bone attached to the hyomandibular above, the symplectic behind, the quadrate below, and the meso-pterygoid in front below.

Meso- or Ento-pterygoid (*Ms.Pt.*).—Also a thin leafy bone very strongly attached to the pterygoid below. Its lower border in front is more strongly calcified than the rest.

Pterygoid or **Ecto-pterygoid** (*Pt.*).—A peculiarly shaped bone. It consists of a strongly calcified piece having the shape of an isosceles triangle, the apex pointing downwards and forwards. When the jaws are shut it is opposed in front to the articular of the mandible. From the posterior angle of its base it sends forwards almost at right-angles a transparent bony rod which is tightly wedged in, and connected by ligament with the meso-pterygoid and the palatine, as in *Sebastolobus*.†

Palatine (*Pa.*).—A curved rod largely of bone, but partly of cartilage. Its posterior half is partly cartilaginous and is closely connected with the meso-pterygoid, pterygoid, and the anterior angle on the base of the pterygoid. Its anterior half is bony except at the extremity, which bears a cartilaginous cap attached by ligament to a dorso-posterior elevation on the maxilla. The palatine sends down opposite the anterior end of the meso-pterygoid a rounded process which is strongly attached to the enlarged anterior extremity of the vomer, and apparently also to the ventral process of the pre-frontal as described by Brooks‡ in the Haddock. In the natural disposition of the bones the palatine lies internal to the maxilla.

Eyeless Side.

Hyomandibular.—Much smaller and less densely calcified, and is altogether an obviously feebler bone, although the ball and socket articulation with the skull, whilst slightly smaller, is yet deeper and stronger. The cartilaginous cap for the inter-hyal is also present on this side.

Symplectic.—Considerably shorter, but more robust, and has only a cartilaginous wedge at its upper extremity.

Quadrate.—Slightly shorter but wider antero-posteriorly and more densely calcified, especially at its free ventral extremity. Its dorso-anterior margin is, however, cartilaginous where it articulates with the meta-pterygoid.

Meta-pterygoid.—Somewhat shorter and narrower,

† Starks, Proc. Californian Acad. Sci., ser. iii., vol. i., 1898.

‡ Sci. Proc. R. Soc., Dublin, iv., 1884.

but the same thin laminate bone. Has only a slight articulation with the meso-pterygoid in front instead of the extensive one of the other side.

Meso-pterygoid.—Not attached to the dorsal edge of the pterygoid but to its inner face. It hence occupies a different plane to that of the meta-pterygoid. It is further much smaller on this side.

Pterygoid.—This is of a different shape on this side. The forward thin rod is here thick and in fact stouter than the remainder of the bone, which is reduced and apparently merged into the forward portion. The left pterygoid is both larger and stouter than the right—thus differing from the left palatine, as will be seen below.

Palatine.—Greatly modified on this side. Its anterior extremity articulates directly with the upper end of the maxilla instead of by the intervention of a short ligament. In one specimen there was also a short ligament containing a sesamoid connecting the same extremity with the anterior forward process of the pre-frontal. Behind the anterior end there is a strong dorsal articulation with the pre-frontal which is not found on the other side. The ligamentous connection with the vomer also exists on this side, but is apparently with the vomer only. The articulation with the pterygoid is also different, the posterior extremity of the palatine being forked and the pterygoid fitting into the split thus formed, the interstices being filled with cartilage. The left palatine is much shorter than the right, but is more robust (cp. the pterygoid).

3.—The Jaw Apparatus (Fig. 5).

Ocular Side.

Articular (*Ar.*).—Has a saddle-shaped articular surface for the quadrate, behind which it sends up a prominent "post-glenoid" process. The bone is stoutest at this region, but becomes gradually lamelliform forwards, and its anterior margin is furcate—the lower limb fitting into the cavity of the dentary. The quadrate facet is cartilaginous. The outer face of the articular is convex, the inner concave.

Angular (*An.*).—A small but perfectly distinct bone situated at the postero-inferior angle of the articular.

Meckel's Cartilage.—A long thin rod of cartilage embedded in the articular behind, but lying quite freely for the greater part of its length. It is situated near and parallel to the ventral border of the inner or concave face of the articular and projects slightly beyond the anterior extremity of the ventral limb of the latter, the free end not being ossified as a mentomeckelian as in *Amia*. The free portion of Meckel's cartilage was 12mm. long in the specimen now described, and in a very large fish it attains a diameter of about 2mm.

Dentary (*D.*).—A thin bone, but well ossified at its dorsal and ventral borders and strongly attached to the dentary of the other side. It is strongly and almost equally forked behind, and contains a large triangular cavity for the reception of the lower limb of the articular. Like the latter, its outer face is convex and the inner concave. Quite near the symphysis it on this side and in this specimen bore 3 teeth, opposite to which on the ventral border a prominent process was set down. In a very large specimen examined there were no teeth on this side, their position being occupied by a roughened ridge.

Maxilla (*Mx.*).—Takes no part in the gape. A stout club-shaped bone, the expanded lower extremity of the handle or shaft of which overlaps the lower jaw externally at about the junction of the articular and dentary when the mouth is closed. The shaft narrows down before expanding to form the head, which is capped with cartilage. The head has 4 articulations: (1) It is closely attached above to the large unpaired intermaxillary cartilage‡ (*I.M.C.*), to which on its other side the left maxilla is also attached; (2) ventrally the maxilla is capped by a moveable piece of cartilage, much smaller and situated ventral to the inter-maxillary. This moveable or gliding cartilage† works in the groove to the right of the beak-like mesethmoid prominence (cp. fig. 1), and also over the large convex facet on the right side of the head of the vomer. It is connected with the inter-maxillary and is excavated on both surfaces to receive the head of the maxilla and the vomerine facet. The latter or free excavation is so contrived as not to interfere with the movement of the cartilage above the vomer. The reader who consults a dried cranium of the Plaice when reading this description will understand that the movement permitted to the maxillary bones by these facets and gliding cartilage is an oblique dorse-ventral one in the direction of the eyeless side (cp. Traquair *op. cit.*). This explains the well-known fact that Plaice are able to pick up food lying on the sea bottom by twisting the mouth towards the lower or eyeless side. In other Pleuronectidæ the twisting of the mouth is towards the ocular side; (3) dorso-posteriorly the maxilla, as already pointed out, is connected by ligament with the anterior extremity of the palatine; (4) anteriorly and externally it is deeply excavated for the reception of the pre-maxilla. The anterior edge of the maxilla is close to and partly overlaps the pre-maxilla.

Pre-Maxilla (*P.Mx.*).—Forms the dorsal part of the gape and consists of two arms. Its vertical arm forms the gape and bore 4 teeth in the specimen now described,

‡ We follow Traquair in using this term for the cartilage in question.

† Cp. Traquair, Trans. Linnean Soc., London, **xxv.**, 1865.

which, however, contrary to those of the opposite side, were placed nearer the posterior edge of the pre-maxilla than the anterior. Dorsally this arm is closely connected with the left pre-maxilla. The posterior arm passes backwards over the inter-maxillary, with which it is closely connected. At the junction of the two arms a process is sent backwards which is both capped with cartilage and is covered by a further loose moveable piece. This process fits into the excavation on the maxilla already described, and hence the pre-maxilla here overlaps the maxilla.

Inter-Maxillary Cartilage (*I.M.C.*).—This cartilage plays an important part in the movement of the jaws. It is asymmetrical, blunt in front but more pointed behind, and forms as it were a pivot on which the two maxillary bones on each side turn. It fits into the depression very obvious in the dried cranium above and to the left of the mesethmoid prominence, and glides up and down from this depression over the prominence itself. Its posterior surface is obliquely grooved, and in such a way that as it moves downwards it passes obliquely over the prominence towards the eyeless side—thus further assisting in the torsion of the jaws to that side.

Eyeless Side.

Articular, Angular and **Meckel's Cartilage**.—The two former are slightly larger than the right, and are also slightly more densely calcified, but the differences are small. Meckel's cartilage was in the specimen now described 2mm. longer on this side.

Dentary.—Appreciably larger and more densely calcified, and is strongly curved whilst the right is almost flat. The forking of the posterior margin is further very unequal, the lower limb being much the larger (cp. fig. 5). The depression at about the middle of the outer face of the articular immediately in front of the quadrate articulation for the M. adductor mandibulæ is much more marked on this side, where the muscle is naturally larger and, further, the distortion of the jaws to the eyeless side is assisted by a tendon from it inserted into the maxilla. The dentary of this side bore 22 teeth, as against 3 on the right side.

Maxilla.—Distinctly larger and more curved than the right but not so robust. At about a third from the head on the posterior edge is an eminence for the attachment of a stout tendon arising in connection with the M. adductor mandibulæ, and the action of which tends to draw the jaws towards the eyeless side. This eminence and tendon are not conspicuous on the ocular side, and indeed in the Sole, where the ocular is also the right side, the tendon is stated by Cunningham† to be absent on the eyeless side, although the muscle is said to be larger on that side. The eminence is figured and the muscle described by Traquair,‡ who calls the latter the Retractor Maxillæ.* At the head of the maxilla on the posterior side the bone articulates directly with the free anterior extremity of the palatine instead of by the interposition of a short ligament. The cap of cartilage gliding over and above the head of the vomer is larger and the terminal free facet is also more extensive. The whole action of the jaw apparatus is markedly asymmetrical owing to the unpaired mesethmoid prominence separating the two maxillary facets being obliquely set towards the right. Hence when the maxillæ are depressed they follow an oblique direction towards the left or eyeless side. The articulation of the maxilla with the pre-maxilla is also modified on this side. On account of the motion of the jaws towards the left the

† The Common Sole. Plymouth, 1890, p. 48. ‡ *Op. cit.*, p. 279, Tab. 30.
   * Cp. also Allis, Jour. Morph., xii., pp. 552 and 576.

head of the maxilla is brought into closer connection with the posterior ascending process of the pre-maxilla. Then again the posterior articular process for the maxilla at the junction of the two arms of the pre-maxilla is smaller, and instead of overlapping the maxilla is overlapped by it.

Pre-Maxilla.—Both arms are longer and stouter. The ascending arm is at right-angles to the oral arm instead of at an obtuse angle as on the right side, and passes over the inter-maxillary more to the middle line of that car tilage. It bore in this specimen 17 teeth as against 4 of the other side, and set, as already stated, in a different plane.

The asymmetry of the suspensory and jaw apparatus, whilst undoubtedly initiated by the torsion cf the cranium, has also been independently emphasized by the habits of the fish, as already described. The broad anatomy of this distortion is as follows: (1) The suspensory apparatus on the right side is mostly longer—thus thrusting the jaws over to the left;† (2) the motion of the maxillæ at the sides of the mesethmoid prominence and over the head of the vomer and the course of the inter-maxillary over the mesethmoid beak itself is an oblique one with a set towards the left, which the pre-maxillæ and mouth must necessarily follow. The jaws themselves and the bones immediately related to them are naturally more robust on the left side, since their function is mostly performed on that side. Hence the practical absence of teeth on the right pre-maxilla and dentary

4.—THE OPERCULAR BONES (Fig. 5).

Ocular Side.

Operculum (*Op.*).—A thin laminate bone containing no cartilage except at the articular cup. It is bifid pos-

---

† Cp. Traquair, *op. cit.*, Tab. 30 and pp. 276-7.

teriorly, the apex of the lower arm overlapping the sub-operculum. Its anterior extremity is greatly strengthened by a median ridge of bone deeply cupped in front and forming a strong ball and socket joint with a process on the posterior margin of the hyomandibular.

Sub-operculum (*S.Op.*).—Described in the Sole by J. T. Cunningham as the "Inter-opercular." A leafy bone thinner than the operculum, sending upwards and backwards a long process behind the bifid margin of the operculum. Ventrally it overlaps a small portion of the inter-operculum. The operculum and sub-operculum support the posterior free margin of the opercular fold, and the characteristic posterior process at the base of the pectoral fin (see fig. 23) is formed by the upper extremity of the sub-operculum and the upper limb of the operculum.

Inter-operculum (*I.Op.*).—The "Sub-opercular" of Cunningham. A thin bone but stouter than the sub-operculum. It stiffens the ventral free margin of the opercular fold. The whole of its dorsal edge lies under the pre-operculum, and at about the middle of this edge there is a depression (and here, as in the operculum, the bone is thickest and most strong), providing a ligamentous articulation with the inter- and epi-hyals at the junction of the two latter—a somewhat similar condition to that found in *Amia*.† The connection of the operculum and inter-operculum (and especially the latter) with the hyoid arch, both apparently common in the bony fishes, confirms the view that these elements are modified branchiostegal rays. The sub-operculum is probably also another.

Pre-operculum (*P.Op.*).—This is usually considered to be a modified lateral line bone, *i.e.*, a bone developed primarily around a portion of the lateral line system, and

† Allis, Jour. Morph., xii. Cp. also Shufeldt, Report U.S. F. C., 1883.

is therefore probably not homologous with the other opercular bones. In the Plaice it is L shaped and above overlaps a portion of the hyomandibular. Its straight anterior margins are closely connected with the hyomandibular, symplectic, and quadrate. Opposite the ventral edge of the hyomandibular it almost completely covers the interhyal. Ventrally it overlaps the posterior edge of the quadrate. It is the stoutest of the opercular bones—its anterior surface especially being strongly calcified.

Eyeless Side.

The relations of the bones are precisely the same, but the following differences in shape, &c., may be noted :—

The operculum is slightly smaller and not quite so strongly calcified, and the sub-operculum, though smaller, is somewhat stiffer than the right. The pre- and interopercula are both distinctly smaller and less calcified, and hence the asymmetry is most marked in the anterior opercular bones.

The bones† which enter more or less into the formation of the entire skull of the Plaice may now be provisionally arranged in the following categories according to their manner of origin :—

A.—Bones formed as ossifications within the primitive cartilaginous brain case of the embryo :—Alisphenoid (*Al.S.*), Basioccipital (*B.O.*), Epiotic (*Ep.O.*), Exoccipital (*Ex.O.*), Mesethmoid (*M.E.*, *M.E.*[1]), Opisthotic (*Op.O.*), Prefrontal (*P.Fr.*), Prootic (*Pr.O.*), Pterotic (*Pt.O.*—less the fused dermal pterotic or squamosal), Sphenotic or Postfrontal (*Sp.O.*—less the fused dermal sphenotic), Supraoccipital (*S.O.*).

---

† The only definite cartilages in the Plaice's skull are the ethmoid cartilage (*Eth.*), inter-maxillary cartilage (*I.M.C.*) and Meckel's cartilage. The remaining cartilage is not of an independent character.

B.—Membrane bones which have become secondarily incorporated into the cranium :—

i. Roof of Skull. Frontal (*Fr.*—less the fused lateral line bone), Parietal (*Par.*).

ii. Mouth. Ossifications in the mucous membrane— Parasphenoid (*Pa.S.*), Vomer (*Vo.*).

C.—Lateral line ossicles and bones formed by the modification of such :—Lachrymal (*Lc.*), Nasal (*R.Na.*), Preoperculum (*P.Op.*), Squamosal (see pterotic—*Pt.O.*), Suborbital chain, Supratemporal chain.

D.—Bones formed either within or in immediate connection with the mandibular arch of the embryo :— Angular (*An.*), Articular (*Ar.*—less the fused lateral line ossicle), Dentary (*D.*—less the fused lateral line bone), Maxilla (*Mx.*), Mesopterygoid (*Ms.Pt.*), Metapterygoid (*M.Pt.*), Palatine (*Pa.*), Premaxilla (*P.Mx.*), Pterygoid (*Pt.*), Quadrate (*Qu.*).

E.—Bones formed in connection with the hyoid arch of the embryo :—

i. By modification of the upper segment of the arch itself—Hyomandibular (*Hm.*), Symplectic (*Sy.*).

ii. By modification of the posterior branchiostegal rays of the arch—Interoperculum (*I.Op.*), Operculum (*Op.*), Suboperculum (*S.Op.*).

F.—Compound bones, *i.e.*, cranial bones to which ossicles or bones originally developed in relation to the system of sensory canals have become secondarily fused to their superficial surface :—Articular (*Ar.*), Dentary (*D.*), Frontal (*Fr.*), Pterotic (*Pt.O.*), Sphenotic (*Sp.O.*).

We shall now proceed to describe the visceral skeleton proper, taking the hyoid arch first and the branchial arches afterwards.

D

5.—HYOID ARCH* (Fig. 6).

Ocular Side.

Uro-hyal (*U.Hy.*).†—A short bony rod but cartilaginous in front and behind, and articulating with the upper hypo-hyals and slightly with the first basi-branchial.

Hypo-hyals (*H.Hy.*).—Two pieces, as in the Cod, partly of bone and partly of cartilage and loosely articulating in the middle line with the same elements of the other side. The upper one is perforated in the middle, thus giving a false impression as to the whereabouts of the suture between the two.

Cerato-hyal (*C.Hy.*).—A stout bar traversed in front (anterior face) by a longitudinal groove, the texture of the bone on each side of which running in different directions, thus seeming to indicate that the cerato-hyal, as well as the hypo- and epi-hyals, is either splitting or has been formed by fusion.

Epi hyals (*Ep.Hy.*).—Also double, the lower piece being entirely cartilaginous and the upper partly so. The suture between the cerato- and upper epi-hyal is obscured by an overgrowth of bone as in *Micropterus*,‡ but may be seen on holding the hyoid bar up to the light.

---

* It is here necessary to explain the precise significance of the prefixes *basi-* and *hypo-* as applied to parts of the visceral skeleton. The term basi- can only be applied to a median unpaired ventral element, and the term hypo- to the pair (*i.e.*, one on each side), immediately succeeding it. Now whilst these three elements may, and often do, exist side by side in any one species, the basi- element may be absent, and a median unpaired ventral piece formed by the two hypo- elements fusing together, the result being a secondary basi-segment. In the latter case the terms basi- and hypo- are synonyms (and are used indifferently) ; in the former, they are not.

† The terms basi-hyal, glosso-hyal, ento-glossal and basi-branchiostegal have also been applied to this bone in Fishes by different authors, and the same terms, or some of them, have been applied to *other* elements in higher Vertebrates. The synonymy is too complex to be discussed here.

‡ Shufeldt, *op. cit.*, p. 819, and fig. 32.

Inter-hyal (*I.Hy.*, figs. 5 and 6).—Possibly a sesa-moid bone developed in the inter-hyal ligament and not homologous with the other segments of the hyoid. This bone is incorrectly called by some authors the stylo-hyal— a term really a synonym of the pharyngo-hyal, represented in most fishes by the hyomandibular. The inter-hyal of the Plaice consists of a central bony rod with cartilaginous extremities articulating with the hyoid and symplectic and inter-operculum, as already indicated. Below the attachment of the inter-hyal to the upper epi-hyal is seen the prominence which is also connected with the inter-operculum.

Branchiostegal Rays (*Br.R.*).—There are the usual 7 of these rays, the first on each side meeting at their free extremities, and being closely bound together by strong fibrous connective tissue, appear to fuse. They are not all attached at the same plane as shown in the figure. The first 3 articulate with the cerato-hyal, the last 4 at about the junction of the 2 epi-hyals. The last, however, is always attached to the upper or bony epi-hyal. The rays have cartilaginous extremities, but otherwise consist of a transparent milky coloured bone. The first two cross and lie under the others in the living state.

Eyeless Side.

The hyoid bar is slightly shorter and not quite so robust. The first pair of apparently fused rays are drawn over to the eyeless side as shown in the figure. The most conspicuous difference is in the branchiostegal rays, which are, except the first, uniformly shorter and not so curved. The length and curve are faithfully represented in the figure.

6.—Branchial Arches (Fig. 7).

Ocular Side.

Basi-branchial I. ($B.Br.^1$).—A triangular bone, with its apex wedged in between the upper hypo-hyals (see fig 6). At its base it articulates with basi-branchial II. and with the hypo-branchials of the first arch, but the latter articulation is not obvious dorsally, the head of the hypo-branchial fitting into a deep lateral socket formed by basi-branchials I. and II. According to Cunningham this bone is in the Sole completely wedged in between the upper hypo-hyals and does not articulate with the first branchial arch at all.

Branchial Arch I.—The epi-branchial ($E.Br.^1$) bears a prominent tubercle on its anterior surface. The pharyngo-branchial ($P.Br.^1$) is a slender bone which articulates with the skull at the ventro-posterior margin of the jugular foramen in the prootic immediately below the hyomandibular cup. This somewhat curious connection with the skull also exists in the Sole according to Cunningham, and in *Sebastolobus* according to Starks.

Basi-branchial II. ($B.Br.^2$).—An hour glass shaped bone articulating in front with the basi- and hypo-branchials of the first arch and behind with basi-branchial III. and slightly with the hypo-branchials of its own arch.

Branchial Arch II.—The hypo-branchial is wide but compressed dorso-ventrally. Where it articulates with the second basi-branchial it sends down a prominent spine. Another well-marked spine is borne on its anterior edge. The cerato-branchial is longitudinally grooved ventrally. As in the first arch, the epi-branchial bears a tuberosity, but it is much more prominent on this arch. The first and last gill rakers are very small. The superior pharyngeal bone of the Plaice in medium-sized fish consists of three pieces, which represent the pharyngo-branchials of the three arches to which they belong. These pieces are, however (and especially the anterior two), so closely bound together that they may, as we have known them to do in other forms, fuse up in old fish. The second pharyngo-branchial ($P.Br.^2$) is a stout laterally compressed bone articulating with the third pharyngo-branchial posteriorly. It bore five teeth in one row.

Basi-branchial III. ($B.Br.^3$). A very thin laterally compressed bone, apparently wedged out of existence by the large hypo-branchials II. Its posterior extremity lies under and is covered by the two hypo-branchials III.

Branchial Arch III.—The hypo-branchial is smaller than in arch II., but the ventral spine is both larger and longer, and articulating strongly with the same spine of the other side forms a bony arch traversed by the ventral aorta. The anterior spine in hypo-branchial II. is absent. The cerato-branchial is grooved ventrally as in arch II., but more deeply. The epi-branchial bears two large tuberosities at its distal extremity. The posterior of these articulates with the pharyngo-branchial III. ($P.Br.^3$), the anterior by two strong ligaments with the epi-branchial IV The pharyngo-branchial ($P.Br.^3$) bears a strong process behind for articulation with the pharyngo-branchial II., and bears eight teeth in two rows.

Basi-branchial IV. ($B.Br.^4$).—A very small nodule of cartilage wedged in between the bases of arches III. and IV. It is only connected with the fourth arch on the ocular side, the basal elements of this arch on either side meeting in the mid-ventral line. The morphological value of this cartilage cannot be determined on adult material. It is obvious that the branchial arches have undergone reduction from behind forwards. Thus there are only three segments in the fourth arch. Now it is also obvious that the basal segment of this arch ($C.Br.^4$) is serially homologous with the cerato-branchials, the missing element therefore being the hypo-branchial. Hence the fourth basi-branchial may represent either the two vestigial hypo-segments fused together (a primary basi-segment being absent) or it may have been formed by the basi- and two hypo- elements fusing up.

Branchial Arch IV —The cerato-branchial ($C.Br.^4$) is slightly grooved ventrally. The epi-branchial ($E.Br.^4$) is a stout L-shaped bone, and is strongly connected with the same segment of the preceding arch. The pharyngo-branchial ($P.Br.^4$) is small, and bore six teeth.

Branchial Arch V.—This is more reduced than any of the other arches, and consists of a single bone on each side in which there are practically no traces of asymmetry. This is the inferior pharyngeal bone of Cuvier, and appears to represent the cerato-branchial segment only of the arch.* The inferior pharyngeals ($I.Ph.$) are stout triangular-shaped bones separate dorsally but bridged in front ventrally by a tract of cartilage. Two irregular rows of teeth are borne on the pharyngeal surface, and in the specimen now described there were 12 on the ocular side and 14 on the eyeless. At the side and at the base of the outer row are situated the replacing teeth, which become functional as their predecessors wear away.

Gill Rakers.— These diminish both in size and number from before backwards. Their function is to provide a rough filtering apparatus for the water passing out of the pharynx. Their small size and number in the Plaice is due to the nature of the fishes' food. In those fishes where the food might easily escape through the gill slits (*e.g. Clupea*), the gill rakers are much longer and more numerous. They are purely dermal and are not fused

* Cp. Cunningham, *op. cit.*, and W. K. Parker, Phil. Trans., 1873 (*Salmo*).

on to the arches, the only connection between the arches and the rakers being that in older specimens and in some places the position of the raker is indicated on the arch by a faint elevation. Their number and position, however, was in the few specimens examined remarkably constant and symmetrical, so that the following formula may apply to either side of most individuals:—

	I.	II.	III.	IV.	V.
Hypo-branchial ...	2	3	0	0	0
Cerato-branchial ...	5	6	7	6	0
Epi-branchial ...	3	1	0	0	0

Gill Rays.—The gill filaments are supported by series of very delicate fragile gill rays fused together by their bases like a comb, which it is hardly practicable to dissect, but which are quite obvious in sections of the gills. They radiate out from the branchial arches as usual, and occur in pairs—one to each demibranch of the arch.

Eyeless Side.

Branchial Arch I.—All segments slightly shorter and not so robust, the hypo-branchial markedly so, nor is the latter so deeply socketed into basi-branchials I. and II. as on the ocular side. The pharyngo-branchial also articulates with the skull, but the depression in the skull with which it is connected is deeper and more marked.

Branchial Arch II.—Practically no difference except that the basi-branchial articulation is stronger on the ocular side.

Branchial Arch III.—The segments are of the same length, but are somewhat less robust. The ventral arch transmitting the ventral aorta has been already described.

Branchial Arch IV.—The cerato-branchial is less strong on this side, but on the other hand the epi-branchial is larger. The cerato-branchial does not articulate with the fourth basi-branchial but with the corresponding segment of the ocular side.

The Superior and Inferior Pharyngeals.—The asymmetry in the number of the teeth is so strongly marked in the mouth, where the ocular side is practically devoid of them, that it is interesting to enquire whether its effect has been felt as far back as the pharyngeal bones. In the inferior pharyngeal the two sides are practically the same, except that in the specimen described the ocular bone bore two less teeth. The left superior pharyngeal, however, was appreciably the larger, and although it only possessed an advantage of one in the number of teeth, the teeth themselves were larger and capable of doing more work. This is doubtless due to the fact that in an animal lying on its left side, its food, even in the pharynx, naturally gravitates to the latter side.

7.—VERTEBRAL COLUMN.*

(Figs. 10, 11, 12, 13, 14, 17, 18 19).

The vertebral column of the Plaice may be divided into a trunk and tail region only, distinguished in the former by the presence of ribs and in the latter by the haemal canal. Although the number of vertebræ in the column is subject to variation, it usually happens that the first caudal vertebra is the fourteenth.

Each vertebra is markedly amphicoelous, the anterior and posterior faces being considerably scooped out in the

---

* The structure and development of the vertebral column of Teleostean fishes has recently been studied by S. Ussow (Bull. Soc. Imp. Nat., Moscou, 1900, p. 175). Also previously in *Amia* and other fishes, by O. P. Hay (Field Columbian Museum, Zool. Ser., vol. i., No. 1, 1895).

form of a cone, the two cones being connected in each vertebra by the pin-hole notochordal canal (Canalis dicentralis). As all these spaces are occupied by the " remains " of the notochord, the latter is absolutely continuous from one end of the column to the other.

The following description is based mostly on a large specimen of an extreme length of 52cm. In this animal the neural spines were inclined as follows: 1, slightly forwards; 2, 3, 4, 5, almost upright; 6, slightly forwards; 7-13, all slightly curved (with the convexity forwards) and project more or less forwards; 14, largest spine (first caudal) and projects slightly backwards; from 1-14 the neural spines increase in length; behind 14 they all incline backwards, the inclination becoming more and more marked as the extremity of the tail is reached, and they also decrease in length. With regard to the haemal spines (of which the anterior ones are very much longer than the corresponding neurals), the first 3 incline slightly forwards; 4 is vertical; 5 looks backwards, and so do the remainder, the tendency becoming gradually exaggerated behind, and at the same time the spines becoming shorter until they are about the same length as the neural spines. In the average specimen the posterior haemals are slightly longer than the neurals (cp. fig. 19).

In the posterior third of the body the vertebral column is situated about half way between its dorsal and ventral edges. In front of this region, partly owing to the slight upward curve of the column, but principally owing to the increased length of the haemal over the neural spines, the column is situated markedly nearer the dorsal than the ventral edge. In the anterior third it begins to bend down again slightly, and this is especially noticeable in the first 4 or 5 vertebræ, the result being that the skull when attached to the vertebral column is directed markedly downwards (see fig. 17).

The anterior notochordal space in most of the trunk vertebræ (except the atlas) is perceptibly deeper than the posterior, so that the notochordal canal is situated nearer the posterior than the anterior face of the centrum. This is more marked in some vertebræ than in others.

Atlas (figs. 10 and 17).—Body compressed from before backwards. Notochordal canal (*N.C.*) much nearer dorsal than ventral surface. Bears two large cartilage capped facets (*C.F.*) for articulating with the paroccipital condyles, of which the left is perceptibly larger than the right. The anterior face of the centrum also articulates with the single occipital condyle on the basi-occipital, the connection of the skull with the vertebral column by means of 3 condyles being therefore very strong. Unlike all the other vertebræ, except about the last 5, the neural arch of the centrum is only perforated by one foramen on each side for the second spinal nerve. There is no trans verse process, and only one. rib, which belongs to the series of accessory ribs or intermuscular bones (*A.R.*[1]), and is attached to its vertebra higher up than any of the others, articulating at the junction of the neural arch with the centrum (figs. 10, 17). As in all the other vertebræ (although the tendency is faint in the posterior caudals), and as first described by Traquair, the atlas is markedly asymmetrical, the neural spine being directed towards the eyeless side. The asymmetry here is obviously an adaptation to the habit of the animal in lying on its eyeless side. Superficially this side is practically flat, whilst the ocular side is convex. The asymmetry of the vertebræ, therefore, tends to a flattening of the eyeless side and an arching of the ocular side (cp. figs. 11 and 13). The neural spine itself is in the form of a rolled plate forming a hollow cylinder, but the edges do not fuse behind. Except that the eyeless condylar facet is larger than the ocular one the asymmetry is not noticeable below the neural spine. The centrum is overlapped by the large ill-defined anterior zygapophyses of the 2nd vertebra, and itself bears faint posterior zygapophyses at the bases of the neural arches, whilst on the ocular side the latter sends back a hook-like process which fits *outside* the neural arch of the second vertebra. This is the only trace of the characteristic method of articulation of the vertebræ of the Cod.

Second Vertebra (fig. 17, $V.^2$).—The neural arch is perforated on each side by *two* foramina for the roots of the third spinal nerve, but the bridge of bone separating the right pair is extremely slender. The large irregular anterior zygapophyses are asymmetrical, that on the eyeless side being much the larger. The neural spine is also asymmetrical as in the atlas, but the asymmetry extends down on to the neural arch. A small pointed transverse process is present, with a long accessory rib ($A.R.^2$) strongly attached to its base much lower down than the attachment of the first intermuscular bone. Both the centrum and neural arch bear post-zygapophysial facets not shown on the eyeless side.

Third Vertebra (fig. 17, $V.^3$).—Whole of the vertebra markedly asymmetrical, being more strongly developed on the eyeless side in every respect. Neural spine and spinal canal arch to the left Anterior and posterior zygapophyses more strongly marked on the same side. The centrum not only bears a strengthening ridge ($S.R.$) which is much stronger on the eyeless side, but is itself more bulky on that side, so that the notochordal canal is eccentric in position. The transverse processes, like all those succeeding them, and as already described by

Traquair (pp. 285-6) are asymmetrical—that on the eyeless side having a more ventral origin and inclination; but the asymmetry of these processes is not specially marked in this vertebra, and may indeed be practically confined to vertebræ 5-12 inclusive. The accessory rib ($A.R.^3$) is attached nearer the base of the transverse process on the eyeless side.

Fourth Vertebra.—Much the same as 3, except: (a) two moderate strengthening ridges are present on the eyeless side, and one on the ocular side with a deep cleft on each side of it; (b) a true rib is present, and is attached to the posterior surface of the transverse process half way between its extremity and the attachment of the accessory rib. From this vertebra onwards the transverse processes increase in length and the true ribs (which rapidly increase in length up to the 7th, the longest [see fig. 18], but rapidly decrease in length after this) gradually approach the tip of the transverse process, until at about the 8th or 9th vertebra they are obliquely attached to the tip. The accessory ribs, which from the 2nd to the 9th inclusive vary very little in length, are also attached in a backwardly descending line until about the 10th or 11th vertebra, after which, just as they begin to decline in size, they rise rapidly on the vertebra (cp. fig. 18), until posterior to the 14th or 1st caudal vertebra they are doubtless represented by the serially homologous tubercles situated on, and fused to, the centrum, and forming pseudo-transverse processes.

Fifth Vertebra.—Neural spine almost flat and only curved backwards slightly at the lateral edges. Asymmetry slightly increasing.

Sixth Vertebra.—Only one strengthening ridge on eyeless side and two on ocular. Asymmetry of centrum and transverse processes strongly marked, but more so anteriorly than posteriorly. The post-zygapophyses project slightly backwards as distinct processes.

Seventh Vertebra.—One strengthening ridge on each side, but slightly cleft into two. Neural spine more com pact and solid. Posterior zygapophyses are now distinct tubercles projecting backwards from about the *middle* of the neural arch. The anterior zygapophyses are, as hitherto, except in the first 3 or 4 vertebræ, triangular processes projecting forwards from the *base* of the neural arch.

Eighth Vertebra (fig. 11).—Two moderate and one weak strengthening ridge on the eyeless side and three of about the same character on ocular side. Posterior zygapophyses as in 7. Anterior zygapophyses much stronger on eyeless side. Neural spine for the most part consists of two hollow tubes placed side by side and connected by a narrow bridge of bone.

Ninth Vertebra.—Two well marked strengthening ridges and rudiment of another on eyeless side and two on ocular side. Zygapophyses getting weaker. Neural arch slightly, and neural spine markedly, asymmetrical, but the centrum is almost symmetrical with the notochordal canal in the centre.

Tenth Vertebra (fig. 18, *V*.10).—Two strengthening ridges on eyeless side and three on ocular. Zygapophyses somewhat reduced, and not much difference between those of the two sides. Centrum slightly asymmetrical and neural and transverse processes still obviously so, but the tendency is now on the wane. The accessory and true ribs of this vertebra are the longest of any. Behind both decline in length.

Eleventh Vertebra (fig. 18).—Three strengthening ridges on eyeless and two on ocular side. Asymmetry slightly less marked than in preceding vertebra. Zygapophyses slightly weaker and more symmetrical. The transverse processes are now increasing in width from before backwards, and bear a slight elevation to which the accessory rib is attached.

Twelfth Vertebra (figs. 12 and 18).—Strengthening ridges as in 11. Asymmetry slight. Post-zygapophyses faint and practically symmetrical. Transverse process proximally very wide from before backwards. Neural spine more consolidated, but still consists of 2 bony tubes placed side by side.

Thirteenth Vertebra (fig. 18).—Three strengthening ridges on eyeless side, and two on centrum and one on transverse process on ocular side. Symmetry as in 12. Transverse processes only very slightly asymmetrical. Accessory rib rudimentary and attached to a prominent elevation near base of transverse process ($A.R.^{13}$). Last vertebra to bear free ribs of either series. Behind the transverse processes and true ribs are converted into the haemal arch and spine. This gradual conversion, and the homology of the parts, is well shown in figure 18. Zygapophyses as in 12. Transverse processes very wide from before backwards.

Fourteenth Vertebra (figs. 13 and 18).—The first caudal vertebra. Three strengthening ridges on each side. The neural and haemal spines are the longest of any, and both incline slightly backwards and towards the eyeless side. The neural spine, which is only about two-thirds the length of the haemal spine, is somewhat complex, and contains three longitudinal cavities, all of which open posteriorly at the top of the spine. The left anterior zygapophysis is much more prominent than the right, whilst the post-zygapophyses are feeble and no longer obvious as projections from the posterior border of the neural arch. The posterior notochordal space, unlike those of the trunk vertebræ, is deeper than the anterior, and hence the notochordal canal is thrown further forwards. The centrum is markedly asymmetrical, being larger on the eyeless side. The projection at the side of the centrum which possibly represents the remains of the accessory rib and its basal tubercle (*A.R.**), with which it is serially homologous and to its present position on the centrum it has been gradually ascending from the transverse process, is larger and has a more ventral inclination on the ocular than on the eyeless side. Below the centrum are situated the haemal arch with its canal and the haemal spine. The latter is a thin curved laminate bone with the concavity directed forwards, and strengthened behind by two thin longitudinal ridges. In front it bears a wide longitudinal recess (*Rec.Ax.*[1]) which lodges the large first axonost of the anal fin.

The following points of interest may be noted in the caudal vertebræ. Behind the fourteenth or first caudal vertebra the vertebral column has a much more uniform structure, except that the first few caudals represent intermediate stages between the characters of the 14th and those of a typical caudal vertebra (cp. figs. 18 and 19). The asymmetry extends right down to the extremity of the column, but is only very slightly developed in the last few caudals and in the epurals and hypurals of the caudal fin.

Neural and Haemal Spines.—As we pass back the spines get more simple in structure, shorter, and project more posteriorly. Both spines are, however, always furrowed longitudinally for the reception of the large vertical sheet of ligament connecting them with each other. The structure of the last vertebra will be described in the section on the caudal fin.

Foramina of Spinal Nerves.—All the neural arches except that of the atlas and usually about the last 5 are perforated by two foramina on each side for the exit of the spinal nerves (figs. 12, 14, 17, 18, 19). The ventral foramen is usually situated a little anterior to the dorsal one. Of the last 5 caudals the last is not perforated at all, and the preceding 4 have only one perforation on each side (fig. 19). The last vertebra, as of course in all the other caudals, has both a neural and a haemal canal.

Zygapophyses.—As shown in fig. 18, the anterior zygapophyses decline after about the 14th vertebra, and about the last 7 caudals have practically no zygapophyses at all. After the 15th the pre-zygapophysis is much reduced, and only very slightly overlaps the vertebra in front or does not do so at all. Hence behind this vertebra the post-zygapophyses are entirely lacking.

Strengthening Ridges.—These are disposed much the same as in the trunk vertebræ, except that in the hinder caudals there is a tendency for the ridges to be collected into one strong ridge situated mid-laterally on the centrum. This, however, does not obtain in all the posterior caudals (cp. fig. 19).

Notochord.—All the caudals have a notochordal canal except occasionally the last. The centra of the most posterior vertebræ are always less ossified than those in front, and hence the notochordal spaces are larger. The urostyle is deeply excavated also, but the excavation extends straight backwards and does not turn up.

Accessory Ribs.— The tubercles which have been identified as the remains of the accessory ribs are well marked on the anterior caudals, but diminish backwards and are practically absent on about the last 12. In the anterior caudals they occupy the position of the transverse processes of the trunk vertebræ.

8.—CAUDAL FIN AND EXTREMITY OF VERTEBRAL COLUMN
(Figs. 14, 15 and 19).

The caudal fin when first dissected appears to be diphycercal, but the asymmetrical articulation of the second hypural bone ($Hp.2$) at once establishes its heterocercal character. Nevertheless the caudal fin of the adult Plaice is one of the most completely masked heterocercal or homocercal fins on record, and is hence of some interest. The termination of the vertebral column is peculiar, as in place of an upturned tapering bony " urostyle " formed by the ossification of the free extremity of the notochord, there is found an expanded fan-shaped plate which, with the second hypural, gives articulation to the greater part of the caudal fin rays. The proximal stout vertebra-like body articulating with the last true vertebra may possibly represent another vertebra, but it is not constricted off in Plaice from 15 mm. upwards, and in the absence of earlier developmental evidence is here described as the base of the urostyle. Dissection of young plaice of a length of 45mm. shows the vertebral column terminating as above described, but similar preparations of still smaller forms of 15-17mm., where the notochord is yet unossified, prove conclusively that the fan-shaped plate is formed by the fusion of the upturned extremity of the notochord with a separate hypural bone (fig. 15, $Hp.3$), and hence the plate ($U. + Hp.3$) of the adult is a compound bone representing the urostyle and a third hypural fused together. It will be noticed in fig. 15 that the free surface for articulation with the fin rays is afforded exclusively by the third hypural—the urostyle taking no part in it. Apart from the independence of the urostyle and third hypural the tail of a form of the above length shows no essential difference from that of an adult, beyond those

E

differences incidental to its stage. The above description should be compared with the apparently similar one in *Sebastolobus* (Starks, *op. cit.*), and with the very dissimilar one occurring in the herring.*

If the last distinct vertebra ($_{fig.}$ 14, *V*.43, 19, *V*.42) be now examined, it will be seen that the neural spine (*N.S*.43, 42) resembles the haemal spine (*H.S*.30, 29) in structure, but that both are peculiar. Each consists of a partly cartilaginous shaft behind and a thin laminate portion in front. The posterior shafts so closely resemble the succeeding epural and hypural bones respectively as to suggest that an epural above and an hypural below have fused on to the laminate portions, which latter are undoubtedly similar to and perhaps represent the neural and haemal spines in front. As, however, we have no positive evidence of such a fusion, the spines in question are here described as simple neural and haemal spines.

Wedged in between $U + Hp.$ 3 and *N.S.* 43 ($_{fig.}$ 14) are two partly cartilaginous spines (*Ep.* 1 and 2) which are closely connected by ligament with each other and with the last neural spine, but which are not sufficiently long proximally to reach the vertebral column. The gradual increase in length of these spines as the animal grows older suggests that they may in senile forms become connected with the vertebral column. That they develop in the same way as the neural spines seems certain, although there are no vertebræ ostensibly belonging to them. One is appreciably larger than the other, and there is a big gap between the second and $U. + Hp.$ 3. They represent the epural or epiural bones of other authors.

The second hypural bone (*Hp.* 2) is of the same shape and structure as the upper piece ($U + Hp.$ 3), both being cartilaginous distally and strongly ossified proximally.

* Duncan Matthews, Fishery Board, Scotland, Report v., 1886.

It articulates as shown in fig. 14 with $U + Hp.$ 3 above and in front, the latter articulation making the two expanded tail bones unequal. The first ostensible hypural ($Hp.$ 1) is a wedge-shaped bone of the same structure as the 2nd, but much smaller. It is closely attached by ligament to the last haemal spine, but proximally does not reach the vertebral column.

Each bone in the tail giving articulation to fin rays bears a thin terminal cartilaginous epiphysis (cp. fig. 14). As in the other fins there is a pad of sub-cartilaginous tissue intercalated between the bones of the vertebral column and the proximal ends of the fin rays. This, as before, is embraced by the diverging halves of the rays ($F.R.$). The number of the latter in the caudal fin varied in the specimens examined around 20—sometimes more and sometimes less. What possibly may be regarded as the typical condition both as regards number and places of articulation is represented as follows (cp. fig. 14):—Shaft of last neural spine, 1; Epural 1, 1; Epural 2, 2; Hypural 3, 6; Hypural 2, 6; Hypural 1, 3; shaft of last haemal spine, 1; total, 20.

Each fin ray is of the same structure as those of the pectoral and pelvic fins, with the exception that there are no articular processes connecting the individual rays with their neighbours, each ray in the caudal fin, therefore, being independent of those immediately above and below it. With the exception of about the three most dorsal and ventral, each ray bifurcates distally, but does not split up further to form a brush as happens in the caudal fin rays of the Sole according to Cunningham.

9.—DORSAL FIN (Figs. 17 and 19).

The dorsal fin commences very far forwards in the pseudo-medial line, and in the specimen on which the present description is based, which measured 52cm., the base of the first dorsal fin ray was only 6mm. from the posterior narial aperture of the eyeless side, *i.e.*, well in front of the cavity of the brain.

As in the paired and caudal fins all the fin rays consist of two pieces. Each fin ray is connected with two further skeletal pieces termed by Cope,* Baur,† and Smith Woodward‡ the baseost and axonost, by T. J. Parker§ pterygiophores, and by Bridge‖ radial elements. The axonost has hitherto been called the interspinous bone, on account of its position between two neural spines. The terms baseost and axonost are adopted in this work, and the precise connections between these two elements and the two pieces forming a fin ray will be described below. In the meantime it may be remarked that usually one or two axonosts are found between two adjacent neural spines in the dorsal and anal fins of the Plaice, whilst the baseost is always situated between and attached to the heads of two adjacent axonosts. The two halves of the fin ray diverge proximally and tightly clasp the baseost (cp. fig. 16). This mechanism was first described by T. J. Parker in *Regalecus*, and has since been described in the Plaice by Bridge. Each axonost, even in old specimens, consists usually of a bony cylinder filled with cartilage. The head is always hollow, and is filled up with a triangular plug of cartilage (fig. 16).

The first dorsal fin ray has no baseost, but its halves diverge as usual and embrace the head of the first axonost with only a small sub-cartilaginous pad between. It is also asymmetrical, the ocular half being slightly the longer.

* American Nat., 1890.   † Jour. Morph., vol. iii.
‡ Catalogue Fossil Fishes, British Museum, and Vertebrate Palæontology.
§ Trans. Zool. Soc., vol. xii.   ‖ Journ. Linnean Soc., vol. xxv.

The axonost lettered 1 + 2 in figure 17 is also asymmetrical, its head inclining to the eyeless side. As this bone supports two fin rays (unlike any other in the body, except the huge axonost 1 of the anal fin), it was carefully examined in a very large plaice, and was there seen to present indications of three pieces, two of which possibly correspond to what would represent the first and second axonosts locked together, whilst the third is the partly cartilaginous proximal shaft wedged in between them and connecting them with the skull. If this interpretation of the first apparent axonost be correct, the first baseost (*Bs.* 1) will be situated between two axonosts as it should be, and not present an anomaly only found elsewhere in the skeleton of the plaice at the anterior extremity of the anal fin. Posteriorly near the head of axonost 1 + 2 is a recess into which fits baseost 1 for the second fin ray. Below this recess is a prominent projection which gives articulation to baseost 2 for the third fin ray. The second fin ray is asymmetrical, the ocular half being the longer.

The third axonost and fin ray are practically symmetrical. The head of this axonost forms a cone, the second baseost articulating in front and the third behind. Below the head the axonost bears leafy projections in front and behind.

In the fourth axonost the leafy projections above are exaggerated and the shaft is reduced to a median ridge on each side of the axonost. This form of the axonost, *i.e.*, with a median spine on each side and thin laminæ in front and behind, represents the typical structure of an axonost, and is admirably adapted for the muscles of the fin ray inserted into it. As described by Cunningham (pp. 47-48) in the Sole, each fin ray has six muscles, of which there is one on each side (the " right and left abductors ") for deflecting the fin ray to the right and left of the median plane and which are not connected with the *shafts* of the axonosts, whilst the other four (the " elevators and depressors ") are arranged 2 on each side, the anterior right and left pair arising from the anterior surface of the fin ray and being inserted into the lamina on the shaft of the axonost in front of the median spine, the posterior pair arising from the posterior surface of the ray and being inserted into the posterior lamina. The latter muscles elevate and depress the fin ray in the median plane.

If the dorsal surface of the cranium cf a plaice be examined, there will be noticed in the pseudo-mesial plane a longitudinal trench-like depression which passes backwards in a slight sigmoid curve from left to right on the left frontal and supraoccipital. This depression (see fig. 1) is connected with the extension forwards of the dorsal fin over the roof of the cranium. Into it fit usually axonosts 6, 7, 8, whilst the axonosts in front of these, though too short to actually reach the cranium, are connected with the depression by means of ligaments. Behind the attachment of the eighth axonost, the occiput suddenly shelves down, and in this depression, *i.e.*, bounded in front by the reduced occipital spine and behind by the first neural spine, are situated the two succeeding axonosts (9 and 10). Behind the 10th the axonosts become related to the neural spines as usual.

The posterior extremity of the dorsal fin is only note worthy in two respects : (*a*) in the presence of 4 axonosts between neural spines 36-37, the largest number found between any two succeeding neural or haemal spines, excepting the anterior extremity of the anal fin ; (*b*) the last fin ray (71) articulates directly with the axonost (70) without the intervention of a baseost. The last four vertebræ have no connection either with the dorsal or anal fins. In the specimen on which the above description was founded the last fin ray was characterized by its sigmoid curve and horizontal position. These features were doubtless anomalies.

10.—ANAL FIN (Figs. 18 and 19).

The mechanism of the fin rays and their skeletal supports is the same as in the dorsal fin. The axonosts are, however, perceptibly longer than those of the latter fin, and they are sometimes called interhaemal bones to distinguish them from the interspinous elements. The structure of the anal axonosts is the same as that of the dorsal ones.

The anterior extremity of the anal fin is interesting in many ways. The posterior boundary of the body cavity is supported by a very stout bone which curves downwards and forwards from its roof, and terminates in a point behind the anus. Above it fits into a deep recess borne on the anterior face of the haemal spine of the first caudal vertebra (fig. 13, *Rec.Ax.*[1]). The ventral point, to an extent indicated in fig. 18 by a ring, is in dead specimens almost invariably found perforating and projecting freely through the skin behind the anus. It seems highly improbable that such a pathological condition can obtain during life, but the skin covering the point must be very thin.* This bone, in the lack of any evidence as to its development, is here called the first axonost, but it is certain that it is more than this. In no other part of the body is a baseost situated anywhere but between two

* We are now certain that the point does *not* perforate the skin during life, but that the latter is somewhat easily ruptured when a plaice is handled and allows the point to protrude. Also that the protrusion in dead specimens is due to contraction following on preservation. To make use, therefore, of this so-called external "anal spine" in classification, as has hitherto been done, is absurd. Since this was written, we note that Kyle arrives at the same conclusion.

axonosts. As therefore the first axonost of the anal fin supports two baseosts completely and another partially, it follows that it has been formed of at least 3 axonosts fused together; but, as before remarked, we have no direct evidence of this.† The first baseost is situated in a depression and the succeeding two on an elevation on the postero-ventral surface of the first axonost, whilst the second axonost is closely opposed for its whole length to the same surface. The axonosts 3 to 7 fit by their tips into a posterior longitudinal furrow borne on the first. It is obvious that, as the first axonost is situated between vertebræ 13 and 14 and axonost 7 lies in front of the haemal spine of the latter vertebra, axonosts 1 to 7 and fin rays 1 to 8 are situated in a space bounded morphologically by two adjacent vertebræ As previously mentioned, in only one other part of the body are as many as 4 axonosts found in a corresponding position.

The posterior extremity of the anal fin presents no features of special interest except that the last fin ray articulates with the last axonost without the intervention of a baseost.

The mechanism of the fin ray has now to be described (see fig. 16). Each ray consists of two longitudinal distally segmented pieces (*F.R. a, b*) held together by a transverse ligament (*F.R. c*). Proximally these two pieces diverge and embrace the baseost (*Bs.*), and also to a limited extent the axonost (*Ax*). The articular surfaces bear pads of a peculiar kind of soft cartilage (*M.C.*). Each half of the fin ray is connected with the baseost by a stout ligament (*F.R. d*). Now the only connection between the

---

† Whatever doubts may arise on this point will be settled by a reference to the condition in *Solea*, as described by Cunningham; and in *Rhombus*, as described by Kyle.

baseost and axonost is a very slight non-ligamentous one, and hence the fin ray and baseost are only held down in the soft cartilage cup at the head of the axonost by the elevator and depressor and right and left abductor muscles (*F.R. e*) of the fin ray. The result is that the fin ray and baseost àre capable of being moved on the axonost in any direction. The axonosts are held in position by two ligaments: (*a*) by a longitudinal vertical ligament which keeps the axonosts at their correct distances from each other and separates the right and left series of the fin-ray muscles; (*b*) by transverse ligaments (*Ax. a*) which keep the axonost from moving from side to side. The head of the axonost contains a triangular plug of typical hyaline cartilage (*Ax. b*) as before mentioned. The whole appa ratus is somewhat asymmetrical as shown in the figure.

The attached table, based on the examination of a 52cm. plaice, has been drawn up to show the number, position, and precise relations of the ribs, neural and haemal spines, and skeleton of the dorsal and anal fins. The specimen had 42 vertebræ (cp. figs. 17, 18 and 19). The division into regions is somewhat arbitrary, since for example the boundary between the cranial and occipital regions is along the axis of fin ray 9 and baseost 8, *i.e.*, *between* axonosts 8 and 9. In the anal fin the axonosts 1 to 7 are situated morphologically between vertebræ 13 and 14 (cp. fig. 18), and hence at this region (as indicated by the line) the correspondence between the two regions of the table has no morphological value. Behind this ambiguous region (*i.e.*, behind and including vertebræ 14) it will be noticed that although there is a very wide disagreement in the disposition of the fin skeleton above and below the vertebral column, yet the numbers of the fin rays are practically the same, *i.e.*, 45 in the dorsal fin and 46 in the anal. Thus, although the physiological result is the same, the means by which it has been arrived at do not exhibit a serial agreement.

11.—PECTORAL GIRDLE AND FIN (Fig. 8).

Clavicle (*Cl.*).—A large curved bone. The upper part or handle is the stoutest and bears a thin lamina behind for the articulation of the post-clavicle. The outer face dorsally is flattened for the reception of the supra-clavicle. Below, the clavicle is connected in the mid-ventral line with the clavicle of the other side by a long symphysis.

Post-clavicle (*P.Cl.*).—A long thin curved bone articulating above by its upper enlarged extremity with the clavicle, and for the rest lying freely in the superficial muscles under the pectoral fin. Its position is partly indicated externally by a scar on the skin. In one specimen examined the post-clavicle was double, the two pieces uniting, however, at the clavicular articulation. According to Cunningham's figure this bone is not present in the Sole.

Supra-clavicle (*S.Cl.*).—A small triangular bone, thin below but stouter above. Ventrally it overlaps the clavicle, and above it is overlapped by the post-temporal. Its upper extremity bears a prominent cartilaginous knob, which fits into a deep pit on the inner face of the post-temporal.

Post-temporal (*P.Tp.*).—This bone, sometimes called the "supra-scapula," differs from the usual Teleostean type in so far as there are only moderate indications of the forking, and there is only one direct articulation with the skull. Above and in front there is a prominent articulation (representing the upper or epiotic limb of the post-temporal) with the pterotic and epiotic. Somewhat below this, and also in front, is a moderate elevation (representing the lower or parotic limb), which is connected by a ligament with the skull at the region of the junction of the pterotic and opisthotic. In the Sole, according to Cunningham, the forking is more marked, and the lower limb is connected with the opisthotic only. The post-temporal in the Plaice is tunnelled by the lateral sensory canal.

Scapula (*Sc.*).—A thin plate, which for some time remains largely cartilaginous, but which is completely ossified in very large fish, having an oblique shelving articulation with the clavicle. Almost one-half of its outer surface lies internal to, and articulates with, the clavicle. It is perforated by the usual scapular fenestra for the R. ventralis of the first spinal nerve.

Coracoid (*Co.*).—Consists of two parts which are, how-ever, continuous: a dorsal part (corresponding to the meso-pre-coracoid of W. K. Parker*), which calcifies late, is thicker than the ventral part, and gives articulation to fin rays; a ventral thin laminate part (the coracoid of Parker) which calcifies early and projects downwards for some distance as a ventral spine. Owing doubtless to the long articulation with the clavicle the connection between these two bones is here a simple and not a shelving one as is the case with the scapula. The ventral part of the coracoid is absent in the Sole according to Cunningham's figure.

Brachial Ossicles.—These are doubtless absent, but may be represented by three structures: (1) a wedge-shaped piece of cartilage attached mostly to the coracoid but partly also to the scapula (cp. fig. 8)—this is present in the Sole according to Cunningham's figure; (2 and 3) two sub-cartilaginous pads, one of which works over the free surface of (1), and the other the free surface of the

* " Shoulder Girdle." Ray Society, 1868.

scapula (cp. fig.). It is these pads that really give articulation to the fin rays, and are in fact interposed between the extremities of the fin rays and the scapulo-coracoid. Judging from the analogy of the other fins there may only be one pad normally present in the pectoral fin (cp. pelvic and caudal fins).

Fin Rays (*F.R.*).—There were eleven fin rays in the pectoral fin of the specimen now described. Each ray consists of two pieces enclosing a central core of soft tissue, as commonly occurs among Teleosts. Each ray is completely segmented for the greater part of its length, so that on maceration it falls into a number of very small pieces. The two portions of the ray diverge at the scapulo-coracoid articulation and embrace its sub-cartilaginous pad as already described. Each portion also sends down an articular process, and the two of each fin ray clasp the ray immediately below it (cp. fig.), thus giving a rigidity to the fin it would not have were the fin rays independent of each other. Three of the rays were bifurcated at their free extremity in this specimen, and where this obtains both halves of the ray split, the bifurcation not being due to the two halves diverging.

"Inter clavicle" (*I.Cl.*).—This bone, of questionable homology, is described last, as it is doubtful what claim it has to the name now given it. It is a median V-shaped bone—one limb of the V being horizontal and the other projecting forwards and downwards. It is situated in the muscular cone of tissue passing forwards to the hyoid arch from the clavicle in the middle line between the basal portions of the branchial arches. The horizontal limb is connected anteriorly by four long stout ligaments with the inner face of the lower hypo-hyal of each side— two ligaments passing to each hypo-hyal. Behind it is connected with the clavicle. The arms of the V are densely calcified, and so is that portion of the lamina at the junction of the arms. The remainder or posterior part of the bone consists of a very thin plate or lamina. The " inter-clavicle " is called by Owen, Huxley and other anatomists sometimes the uro-hyal† and sometimes the basi-branchiostegal, and it is asserted by Cunningham that these names cannot correctly be applied to it. The term (*i.e.*, " jugular ") used by Cunningham is, however, itself inadmissible, since it is liable to be confounded with the jugal or with the jugular plates of " Ganoids "—with the former of which it can have no possible connection. In the Sole, according to this author, it is applied directly to the clavicle and first basi-branchial, and in this differs markedly from the Plaice, where it is placed some distance from both these bones. Its position in the Plaice relative to that of the clavicle is correctly indicated in fig. 8, and it is somewhat further removed from the hyoid arch. Hence, whatever its position in other Teleosts, in the Plaice it is directly connected neither with the clavicle nor with the hyoid. In *Sebastolobus* according to Starks, and in *Micropterus* according to Shufeldt, what seems to be the undoubted homologue of the " inter-clavicle " articulates with the hyoid arch and is called by these authors the " uro-hyal." Cunningham's objection to this term, however, seems to be valid, and hence the provisional name of " inter-clavicle "—a bone with which it may not unreasonably be identified. On the other hand its connection in other forms with the hyoid arch indicates that it may have been derived from branchiostegal rays, in which case it may conceivably be homologous with the jugular

† The supposed resemblance to the uro-hyal of the bird doubtless suggested this homology. Kyle has recently revived this name for the bone, but does not seem to be aware that the terms uro- and glosso-hyal are too often used as synonyms to justify any separation now. In any case, we consider the term uro-hyal quite inadmissible.

or gular plate of *Amia*. This, however, is so very problematical that to prevent confusion, Cunningham's term must be rejected.

12.—Pelvic Girdle and Fin (Fig 9).

Innominate or Pubic Bone (*In*).—Situated just behind the ventral extremity of the clavicle. It is strongly connected by its dorsal extremity with the inner face of the clavicle opposite the upper boundary of the clavicular symphysis. It is also connected more or less for the whole of its length, and especially ventrally, with the innominate of the other side. In medium-sized plaice it consists of three pieces, but the tendency is for the lower extremity to calcify. The dorsal piece (1) is the largest and consists of a fairly strong posterior rod, produced below into a ventral spine, to which is attached in front a thin bony plate. Ventrally there is an obvious piece of cartilage (2) which gives attachment to a terminal cartilaginous epiphysis (3), which in its turn supplies the surface over which works the sub-cartilaginous pad giving articulation to the fin rays.

Fin Rays (*F.R.*).—Articulate directly with the innominate except for the intervention of a sub-cartilaginous pad as in the case of the pectoral fin. The fin rays here resemble those of the pectoral fin, and consist each of two pieces, but the latter diverge more proximally, and the posterior articular processes are terminal instead of sub-terminal as in the pectoral fin. The fourth and fifth of these processes, further, projected backwards and downwards instead of straight backwards as with the others (cp. fig.). None of the fin rays bifurcated, and there were the normal six of them in the pelvic fin of the specimen now described.

C.—THE BODY-CAVITY AND ITS VISCERA.

We propose describing under this section the alimentary canal, the digestive and ductless glands and the renal organs, leaving the reproductive organs to be described in Section G.

1.—THE CŒLOMIC SPACES.

The derivatives of the embryonic cœlom are (1) the body cavity, (2) the pericardium, (3) the cavities of the ovaries (in the female), and (4) the cavity of the ureters. The body cavity is bounded dorsally by the kidney, which lies underneath vertebræ 2 to 14, anteriorly by the posterior fibrous wall of the pericardium, the lower portions of the clavicles, the innominate bones and the muscles of both limb girdles, and posteriorly by the 1st haemal spine (*H.S.* 1, fig. 21) and the 1st axonost (1. *Ax.*). It contracts ventrally, so that only a small region surrounding the anus is bounded by the ventral body wall. The lateral body walls are strongly muscular, and the parietal peritoneum is deeply pigmented. There is no posterior extension of the body cavity on either side, such as occurs in the Sole, and is stated by Kyle to exist also in the Plaice.

The nature of the cavities of the ovaries and renal organs is best considered with the description of those organs.

2.—THE ALIMENTARY CANAL AND ITS GLANDS.

The alimentary canal may be conveniently divided into the following regions: œsophagus, stomach, duodenum, intestine and rectum. Œsophagus and stomach are distinguished from each other and from the rest of the alimentary canal by the differentiation of the mucous membrane. The duodenum is the proximal portion of the post-pyloric intestine which bears the pyloric cæca and receives the bile and pancreatic ducts. There is no essential difference between intestine and rectum, but it is convenient to distinguish the terminal portion of the alimentary canal from that immediately preceding it.

The appearance presented by the viscera on opening the body cavity of a large plaice from the ocular side varies considerably with the condition of the reproductive organs and with the sex. Fig. 20, pl. V., represents the relations of the viscera in a " ripe " and mature female. The great increase in volume of the ovaries has crowded the greater portion of the intestine towards the dorsal part of the body cavity; the duodenum is pressed forward; the rectum being more fixed than the rest of the post-pyloric intestine is not much displaced : the stomach occupies nearly its normal position. Fig. 21 shows the condition of the viscera in a mature female which has just spawned. The greater portion of the intestine has been removed, however, in order to display the deeper viscera. It would form two S-shaped loops overlying and hiding most of the structures indicated in the figure.

The **Œsophagus** is very short, and almost immediately on entering the body cavity expands into the stomach. Its walls are very thick and are composed almost entirely of a transversely disposed layer of striated muscle fibres. The external longitudinal muscle layer is thin, and appears to consist of unstriated fibres. The mucosa consists of a layer of columnar cells crowded with " goblet " cells. As observed in the dead fish, the lumen of the œsophagus is greatly reduced, though it is evident from the nature of the food that it is capable of considerable expansion.

The layer known as the muscularis mucosæ does **not**

appear to be present in the intestine of the plaice. Nor is the peculiar " stratum compactum " of the submucosa which Oppel has described in some other fishes certainly present.

The **Stomach** (fig. 21) is sharply distinguished from œsophagus and duodenum by the strongly developed transverse musculature at its proximal and distal ends. The transverse muscle layer is less strongly and the longitudinal layer more strongly developed than in the œsophagus. At its pyloric end the transverse muscle layer becomes much thicker and forms the prominent sphincter pylori, a valve which projects into the cavity of the duodenum. There is also a very marked differentiation of the mucosa. In the œsophagus this consists of a simple columnar epithelium with goblet cells. In the stomach the goblet cells disappear and the epithelium is evaginated to form a closely-set series of gastric glands over the whole internal surface. Each gland is a tubule, the internal portion of which is straight and the deeper portion convoluted. The straight or conducting portion has a wall consisting of columnar cells with a cement substance between them, and the lumen is relatively wide. The deeper or secreting portion has walls made up of large cubical clear cells, whilst the lumen is narrow. ' The submucosa consists of loose areolar tissue containing blood vessels. The stomach lies along the dorsal wall of the body cavity, and the pylorus is situated at the posterior end of the kidney.

The **Duodenum** lies along the posterior wall of the body cavity towards the eyeless side of the body. Its proximal end is slightly folded over the distal end of the stomach. Its wall (and that of the succeeding regions of the alimentary canal) is thin and consists of an outer longitudinal and an inner transverse layer of unstriated muscle fibres, of a sub-mucosa lying internal to these, and of a mucosa which is a simple columnar or cubical epithelium containing goblet cells. Four pyloric cæca (Cœ. fig. 21) are present. Day ("British Fishes," vol. II., p. 26) states that only two are present, and Kyle apparently also agrees with this. When the intestine is distended with food these cæca may be obscured, but their presence may always be determined, and in the young fish (of 2 to 4 inches long) they are usually particularly noticeable. One is present on the dorsal and proximal extremity of the duodenum, two on the ventral and proximal extremity, and one on the mid-ventral line about an inch distant (in large fishes) from the pylorus. This last cæcum may be the largest of the four. They have a wide lumen freely communicating with that of the duodenum; their wall is very similar in structure except that the muscle layers may not be so distinct.

The **Pyloric Cæca** are only seen in Teleostomatous fishes. The number present is very variable, none being found in the Sole and 191 having been counted in *Scomber*. There has been much discussion as to their morphology and function. At one time they were regarded as the homologues of the pancreas—an organ which was then supposed to be absent in Teleostomi. They have been regarded as absorptive organs and as accessory digestive glands. Mordecai from observations on *Clupea sapidissima* supposed that they served to store up reserve food material. In the fishes ascending rivers to spawn, when presumably no food was being taken, the cæca were found distended with a brownish mucus-like substance which was absent at other times in the year. Edinger supposed them to exercise an absorptive function. Wiedersheim also held this opinion, and correlated their presence with the absence of a spiral valve (a device for increasing the absorptive surface of the intestine). Many investigations have been made on their power of secreting enzymes, and the results obtained are confusing. Macallum* investigated the structures in *Acipenser*, taking particular care to avoid the entrance of enzymes from the alimentary canal and pancreas, and found no certain evidence of a digestive action of their secretion on starch or proteid. Bonduoy† obtained the opposite results, finding the secretion of the cæca in many Teleosts to behave like trypsin and to act strongly on starch and proteid. Macallum supposes that they represent the remains of a former series of digestive diverticula of the alimentary canal. These became restricted in most vertebrata to certain regions forming the digestive glands of the canal, but a variable number, however, persisted in Teleostomi as the pyloric cæca.

The remaining portion of the intestine lies on the ocular side of the body cavity. The duodenum passes into a tract of intestine which lies along the ventral and anterior walls of the body cavity to the left of the rectum. This is thrown into two S-shaped loops which terminate in the rectum. Near the anus the muscle layer becomes thicker, and the terminal portion of the rectum is also connected to the adjacent body wall by strands of connective tissue.

The **Mesenteries** are difficult to study on account of the convolutions of the intestine. They are best examined in a specimen well hardened with spirit. Two mesenteric sheets appear to be present, though these may possibly represent a single structure. One takes origin from the dorsal and posterior walls of the body cavity in the

---

* Jour. Anat. and Phys., vol. xx., pp. 604-636, 1886.
† Arch. Zool. Exper., vii., pp. 419-460, 1899.

middle lines of the latter, and suspends the stomach and the greater portion of the intestine, but not the duodenum. It is curious that the anterior two-thirds of the stomach are attached to this mesenteric sheet along the mid-dorsal line, but towards the latter third the attachment is on the right side, as if the stomach had been longitudinally rotated from left to right. This mesentery is of course a double sheet of membrane which encloses the urocyst and ureter. A second, apparently distinct mesentery takes origin over the internal surface of the liver, and is attached to the duodenum and to the greater portion of the succeeding intestine. The latter is therefore attached to other parts by means of two mesenteric sheets. The second mesentery described above covers over the spleen and bile duct.

The **Liver** (figs. 20 and 21) is asymmetrical. It consists of two lobes connected by an anterior isthmus of hepatic tissue. The larger of these lobes forms a flat cake lying on the eyeless side of the body cavity, and the smaller lies in the anterior and dorsal corner of the ocular side, its anterior surface being in contact with the posterior wall of the pericardium (*Per.* fig. 20). The organ is suspended to the body cavity wall by the two hepatic veins (*V. hep.* fig. 21) which penetrate the posterior wall of the pericardium, and by a fibrous sheet passing between these and attaching the pericardial septum to the anterior surface of the liver. This anterior surface, as well as the lateral, is smooth, but the internal surface on the other hand is thrown into lobules (fig. 21) by deep furrows in which the factors of the hepatic portal system run, and along which they can be traced for considerable distances.

The **Gall Bladder** (figs. 20 and 21) lies wedged in between the right hepatic lobe, the right surface of the stomach and the right and dorsal surface of the spleen. It is a large thin-walled sac about one inch in diameter in large plaice. It is not imbedded in the liver in any way, and is attached to the latter organ by means of the bile duct only. Its posterior wall is thickened by a little nodular swelling. Its efferent duct leaves the anterior and ventral surface, turns back and runs on the internal surface of the right hepatic lobe partially imbedded in the tissue of the latter. Three groups of hepatic ducts enter it: one of these is situated near the proximal end of the duct, the other two are placed about midway on its course and enter it from opposite sides. Each is a group of three or four ducts. The cystic duct is the portion of the whole duct between the gall bladder and the opening of the first hepatic duct, the remaining portion is the common bile duct. The walls of the bile duct are slightly iridescent, the distal extremity is thick and swollen, but encloses a very narrow lumen. It enters the duodenum between the paired pyloric cæca. Its opening into the duodenum is extremely small, and is very difficult to observe from the interior of the latter. The bile is a transparent slightly greenish fluid. The liver in the plaice, as in all other vertebrates, has a double blood supply, receiving blood from the veins of the intestine by the hepatic portal channels and from the dorsal aorta *via* the cœliaco-mesenteric artery by the very small hepatic artery (*A. hep.* fig. 22). Of these sources the hepatic portal system of veins is by far the most important, and the system of intra-hepatic vessels containing venous blood is exceedingly striking in sections of the organ. The liver is essentially a tubular gland, but the hepatic tissue is disposed in strings of cells in which as a rule the lumen or bile capillary is only apparent from the radiate arrangement of the hepatic cells in transverse section of the strings. Within the liver pancreatic tissue surrounding the bloodvessels is also present.

The **Pancreas** is apparently absent in *Pleuronectes*, but is really present in a diffuse form. It will be noted that a perivascular tissue is stated to occur round all the visceral blood vessels and particularly round the factors of the hepatic portal system. This tissue appears on naked eye examination as a band on either side of the vessel equal to or greater than the diameter of the latter. It lies of course within the thickness of the mesentery in which the blood vessels travel. It is particularly abundant round the vessels in the vicinity of the pyloric cæca, where it forms nodular masses which are also associated with fatty tissue. This diffuse perivascular tissue is the pancreas, which nowhere has the massive form characteristic of Elasmobranch fishes. It recalls the extended form of organ characteristic of Mammalia, except that here the gland acini have a constant association with the blood vessels, forming an investment round the latter. No proper pancreatic blood vessels are accordingly present. Text-fig. 1, B. is a section through a small portion of the mesentery including two branches of the cœliaco-mesenterio artery and a small factor of the hepatic portal system. It will be seen that the mesentery is greatly thickened round the blood vessels, and this thickening is really due to a mass of gland acini. No attempt is made in the figure to represent the structure of these acini, but their radiate grouping round the portal vein is indicated. Elsewhere in this section they had no definite arrangement. Nodular masses of pancreatic tissue are also present round the paired pyloric cæca, and from these some small ducts enter the cæca. Probably a number of such efferent ducts are present, but we have not determined this exactly. The gland acini have the structure characteristic of the

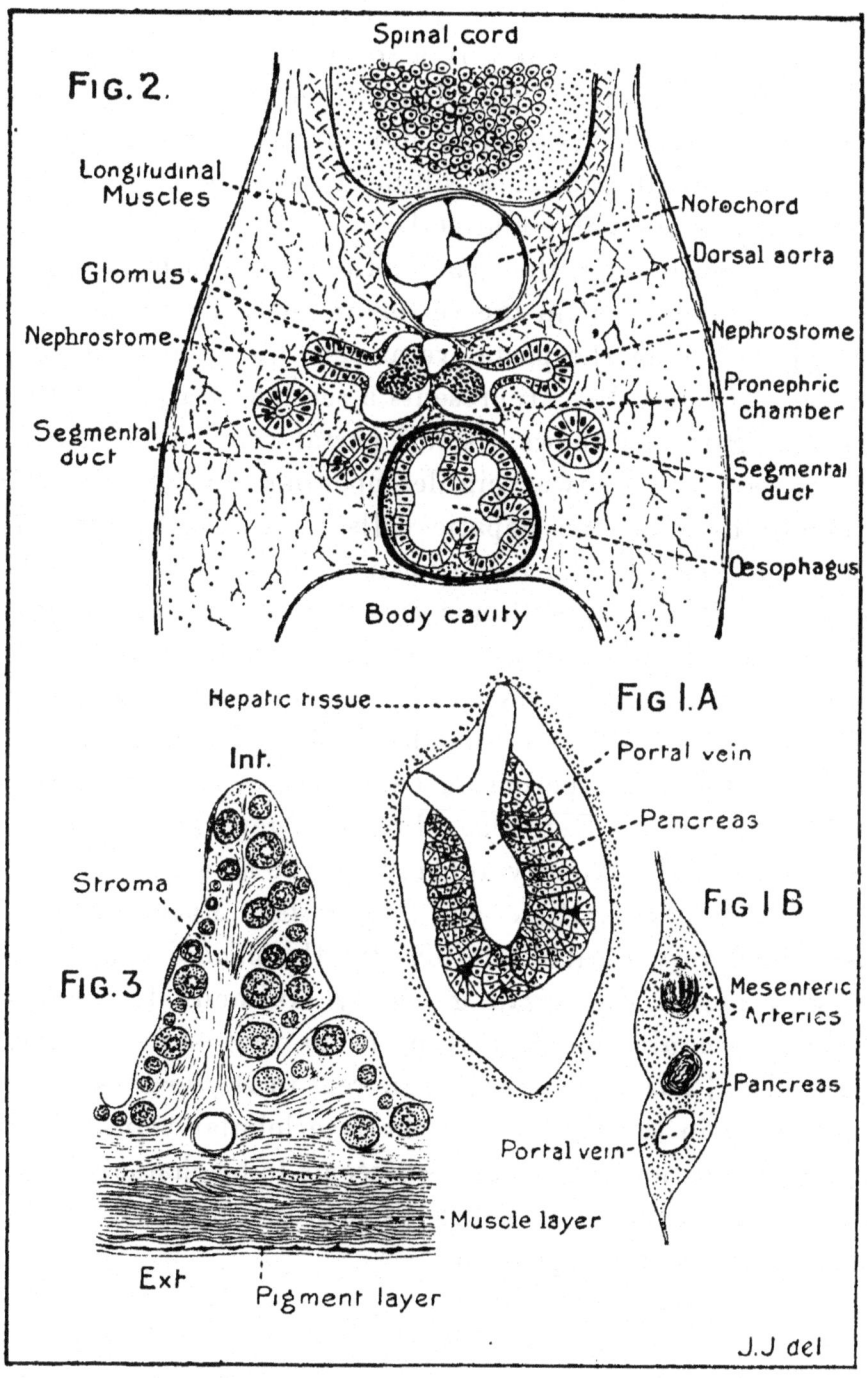

FIG. 2.

Spinal cord

Longitudinal Muscles

Glomus.

Nephrostome.

Segmental duct

Notochord

Dorsal aorta

Nephrostome

Pronephric chamber

Segmental duct

Oesophagus

Body cavity

Hepatic tissue

Fig I.A

Int.

Portal vein

Pancreas

Stroma

Fig I B

FIG. 3

Mesenteric Arteries

Pancreas

Portal vein

Muscle layer

Ext

Pigment layer

J.J del

FIG. 1. A.   Section of liver, showing pancreas round a vein, × 225.
FIG. 1. B.   Section of a part of the mesentery, × 22.
FIG. 2.   Part of a transverse section of a 12 days' old larva.
FIG. 3.   Transverse section of ovarian wall, × 26.

pancreas. A lumen is not generally apparent, but the presence of such is generally associated with a particular phase of the activity of the gland which doubtless did not coincide with the fixation of our material of the organ. They are small and in a single transverse section are composed of few polygonal cells.

Not only does the pancreatic tissue form a perivascular investment in the vessels in the body cavity, but it extends along the portal veins into the interior of the liver. Text-fig. 1, A. represents a section of a part of the latter organ, and shews a small portal vein cut in transverse section with two veinules opening out from it and passing between the hepatic cells. A single layer of pancreatic gland acini forms an investment for the vessel, and the whole lies within a space in the hepatic tissue which is probably natural. The acini are elongated perpendicularly to the surface of the vessel, and the whole is surrounded by a fibrous sheath which sends in partitions between the acini, becoming continuous with the fibrous wall of the vein. Here also a lumen is generally wanting, or is only with difficulty apparent in the acini.

In *Acipenser, Amia* and *Lepidosteus* Macallum* has described very similar relations for the pancreas, and the description of the organ given by Gulland† for *Salmo* answers in all essential respects to that stated above. Macallum‡ has described the pancreas in *Amiurus* as being imbedded in the liver round the interlobular veins. The diffuse condition of the pancreas seems to be characteristic of most Teleostomatous fishes hitherto investigated, and is most probably quite general.

---

* *Loc. cit.*

† Life history of the Salmon ; Rep. to the Fishery Board of Scotland, 1898

‡ Proc. Canadian Institute, N.S., vol. ii., No. 3, pp. 387-417, 1884

3.—THE DUCTLESS GLANDS.

It will be most convenient to consider here a group of glands (the thyroid, thymus, spleen and suprarenal bodies), though these structures have a widely different morphological significance and have most probably very different functions. They agree in being glandular bodies devoid of efferent ducts, and acting in modifying the composition of the blood either by adding to it some substance (internal secretion), or by withdrawing some portion of its constituents. The lymphatic portion of the kidney is also supposed to function in some such way, but this structure will be most conveniently considered together with the renal organs.

The **Thyroid** in *Pleuronectes* is not a compact gland, and is relatively very small in mass. It consists of a number of separate alveoli situated along the course of the ventral aorta. It is difficult to find by dissection in the full-grown specimen, and must be identified in a fish sufficiently small to section as a whole, or by microscopic examination of the tissues surrounding the vessel in question. The separate alveoli of which it is composed are not bound together in any way, but lie loosely in the connective and fatty tissue in which the ventral aorta is imbedded. They are most abundant in the immediate neighbourhood of the origin of the 1st and 2nd efferent branchial vessels, and lie mostly ventral to the aorta, but may be found lateral, and even dorsal, to it. In a transverse section through the region indicated in a small fish $1\frac{1}{2}$ to 2 inches long there may be about a dozen alveoli present in a single section. Round the ventral aorta between the places of origin of the branches referred to and between the 3rd and 4th vessels few alveoli are present, though one or two may be found here and there.

Each alveolus is a small rounded or oval closed sac: the largest is about 0·07mm. in diameter, but they vary in diameter within wide limits. The wall is made up of a single layer of columnar cells outside which a delicate sheath may often be distinguished. Each alveolus is filled with the colloidal substance characteristic of the thyroid, which usually stains slightly with eosin. It may be contracted away from the wall in section, but this appearance is most probably artificial and in life the alveolus is undoubtedly filled.

The thyroid originates in *Salmo*[*] (and probably in all Teleosts) as a median evagination of the ventral pharyngeal epithelium which has no connection with the gill clefts. This evagination forms a little vesicle, the cavity of which at first communicates with that of the pharynx by a tubular stalk. Later on the stalk becomes solid and the vesicle separates entirely from the pharyngeal wall. It then shifts backwards towards the heart, and its wall begins to form hollow buds, which later on separate and become closed. These persist as the definitive thyroid alveoli. Paired rudiments do not, as in higher vertebrates, contribute to the formation of the adult gland.

The **Thymus** (*Thm.* fig. 21) lies internal and slightly posterior to the posterior and dorsal corner of the operculum. On dissecting away the skin in this region a sheet of muscle fibres is seen originating at the cranial ridge connecting the 4th and 5th tuberosities (*Tb.* 4; *Tb.* 5) and inserted into the dorsal border of the operculum. When this is dissected off the thymus is seen lying underneath and immediately in front of the supra-clavicle (*S.Cl.* fig. 8), between this and a slip of muscle which originates in the pterotic at the base of the 4th

---

[*] Maurer, Schilddrüse u. Thymus der Teleostier. Morph. Jahrb., Bd. xi., pp. 130-175, 1885.

tuberosity and is inserted into the upper end of the clavicle. It is a flattened glandular mass occasionally covered with black pigment spots. In a fish of about 20 inches in total length it is about 1½cm. in length, half that in breadth, and about 2mm. in thickness. It lies with one edge uppermost. Its artery appears to be a branch of the subclavian, its vein opens into the superior jugular. It lies immediately external to the roots of the vagus, and this association of the gland and nerve appears to be a constant one in fish of all sizes from the stage at which the structure is definitely formed. Its histological structure is that generally characteristic of the thymus gland of vertebrata. It consists of small rounded cells closely packed together in a narrow-meshed reticulum of connective tissue. The cells have large nuclei and attenuated cell bodies. The whole gland is surrounded by a loose capsule of connective tissue which is continuous with the internal reticulum. There is no well marked differentiation into cortical and medullary regions except that in adult specimens a narrow peripheral zone stains more intensely than the rest of the gland. Fatty tissue is little developed, and concentric corpuscles are apparently absent.

The thymus in *Salmo** (and probably in all Teleostean Fishes) develops, like the organ in Elasmobranchs, from proliferations of the epithelium clothing the dorsal extremities of all the gill clefts. These originally separate thymus buds fuse together while still in connection with the gill clefts, and for a time their cells can be traced continuously into the epithelium of the cleft. The whole organ then separates from its parent tissue and undergoes a backward shifting into its adult position.

The **Spleen** (*Spl.* fig. 21) lies on the internal surface of the left lobe of the liver, usually wedged in between

* Maurer, *Loc. cit.*

the pylorus, the lower surface of the stomach and the distended gall bladder. The bile duct (*Bd.*) lies immediately underneath it. It is covered over or attached by the mesenteric sheet connecting the duodenum and succeeding portion of the intestine to the liver. A prominent branch of the cœliaco-mesenteric artery (*A.sp*. fig. 22) supplies it with blood. It is oval in shape, and in a large fish is about 1 to 2cm. in longest diameter. It is black in colour, and is very soft and easily torn. Its structure is the characteristic splenic one. There is a strong fibrous investment which is continuous with strong trabeculæ passing inwards and dividing the whole gland up into lobules. The reticular formation within the lobules is filled with the characteristic splenic pulp, and in the centre of each lobule is a mass of densely aggregated lymphoid cells round which the texture of the pulp is looser. A prominent vessel passes to each of these nodules. This structure is best seen in the organ of quite small fish, as in larger specimens it is much more obscure, and granular masses of black pigment are abundantly present. These pigment masses are composed of rounded granules of variable diameter.

The **Supra-renal Bodies** are situated on the morphological dorsal (spinal) surface of the kidney on the perpendicular surface which is apposed to the 1st haemal spine at about ⅓ of the length of this surface from the extreme tip of the kidney. To display them the kidney must be removed from the body, and since this is difficult on account of the soft pulpy nature of the organ in the fresh condition, the dissection is most conveniently carried out on preserved specimens. The structures in question are then seen as two oval or rounded bodies lying close together, one on each side of the middle line to right and left of the common genital artery (*A.gen.*) at the point where the latter vessel leaves the kidney. They are yellow or pink in colour, and contrast strongly with the pigmented kidney. In a specimen of about 22 inches in total body length, the largest measured 5·5mm. in longest diameter. They lie in little cavities in the kidney tissue, but project slightly above the surface of the latter, in the capsule of which they are enclosed. Their blood supply is from a twig of the common genital artery, and their minute structure is somewhat similar to that of a lymphatic gland. The capsule is continuous with a system of fibrous trabeculæ traversing the whole organ. These trabeculæ form the coarser bars of a reticulum the meshes of which are crowded with small cells which may be described as lymphoid in appearance. According to Vincent they are secreting glands affecting the composition of the blood by furnishing an internal secretion.

Until comparatively recent times it was supposed that the suprarenal bodies were absent in Teleostean fishes.[+] It has, however, been shewn by Vincent[*] that they are probably universally present. As is well known, the suprarenals of mammalia consist of two morphologically distinct portions—the cortex and medulla. The latter has been stated to have been derived from certain of the sympathetic ganglia, and concerning the former an interesting suggestion has been made by Weldon and Grosglik. It is certain, however, that the most important relations of the suprarenal bodies are with the vascular, not the nervous system.

4.—The Renal Organs.

The **Kidney** (figs. 20 and 21) lies along the whole dorsal, and part of the posterior wall of the body cavity,

[+] Cp. Weldon—Head Kidney in Bdellostoma—Studies Morph. Lab. Cambridge, vol. ii., pt. 1, 1884.

[*] Contributions to Anatomy and Histology of the Suprarenal Capsules. Trans. Zool. Soc., London, vol. xiv., pp. 41-84, 1896-8.

from the 2nd to the 14th vertebræ. Its greater portion is covered laterally by the transverse processes and ribs of those vertebræ. Its ventral surface forms the roof of the body cavity. Posteriorly it increases very much in thickness dorso-ventrally and occupies the angle formed by the vertebral column and the nearly perpendicular 1st haemal spine and axonost, and laterally by the 7th to 10th ribs. It is for the greater portion of its length a single undivided mass, but anteriorly is produced into two tapering and diverging horns—the head portions of the kidney, which lie laterally and dorsally from the œsophagus. The right unpaired cardinal vein runs along the middle line as far as the thickened portion, and is visible on its ventral surface. Dorsally the aorta lies in a groove in the middle line, and this with the cardinal vein separates the uriniferous tubular tissue into two paired masses. Only in the thickened posterior portion of the kidney is this tubular tissue continuous across its whole breadth.

At the dorsal posterior corner of the kidney the caudal vein enters as a single vessel which almost immediately divides into paired portions. Apparently it does not become continuous with the cardinal vein, but breaks up round the uriniferous tubules, though there are doubtless anastomoses between the two vessels. The extreme ventral portion of the kidney is produced downwards into paired tips, and into these the paired genital veins (*V.gen.* fig. 21) enter. Along the dorsal surface of the organ other paired venous trunks (parietal veins) also enter. The most anterior of these vessels is shewn in fig. 22 entering the extreme anterior tip of the head portion of the kidney. The caudal, genital and parietal veins, with the renal arteries are the afferent vessels of the kidney. The paired posterior cardinal veins are its efferent vessels.

The **Ureter** (*Uret.* fig. 21) leaves the ventral and posterior surface of the kidney between the paired terminal processes into which the genital veins open. It is a single tube which immediately on entering the kidney divides to form the paired segmental (or Wolffian) ducts which traverse the entire length of the organ. The ureter rapidly expands into the urocyst (urinary bladder), a large thin walled sac lying between the ovaries (or testes) in front of and rather to one side of axonost 1. Its most expanded portion is near the rectum. Its cavity then rapidly diminishes, and the efferent ureter is a tube with an extremely contracted lumen. It passes to the right side and runs forwards in the dense connective tissue of the body wall, surrounding and posterior to the anus. It then curves laterally at a sharp angle and opens externally on to the surface through the urinary papilla. The latter (*Ur. pp.*) is an unpaired prominent projection of the body wall to the right side and immediately posterior to the anus.

Detailed observations of the development of the Teleostean urocyst are few, but it seems most probable that it is of hypoblastic origin, and that its cavity, unlike that of the segmental duct, which is cœlomic, is really a cloacal portion of the hind gut. McIntosh and Prince[*] give a description of the early condition of the vesicle in *Molva*, though its origin or latter fate is not described. In the youngest plaice (one week after hatching) of which we have made serial sections the united segmental ducts appear to open into the hind gut, which does not yet open to the exterior. In plaice a fortnight old the hind gut opens externally, and the urocyst ceases to have any connection with its cavity, but opens independently in the middle ventral line of the body immediately behind the anus.

The secretory tissues of the kidney are the uriniferous

* Development and Life Histories of Teleostean Fishes. Trans. Roy. Soc., Edinburgh, vol. xxxv., pt. 3, No. 19.

tubules, which are somewhat sparsely distributed in the anterior part of the organ, but are very abundant in the thickened posterior portion. They are greatly convoluted tubules of varying diameter opening into the ureters. It is somewhat remarkable that Malpighian bodies are very difficult to find, and indeed seem to be absent, in some parts at least, of the kidney. This condition is connected with the vascular arrangements of the organ. By far the greater portion of the blood entering it is venous, and the arterial supply is very scanty. Two, or at most three, very small vessels originating directly in the dorsal aorta enter at the dorsal surface, and the common genital artery (*A. gen.*, fig. 22) gives off several very fine arterial twigs which ramify in the posterior portion. The whole arterial blood supply is very small compared with the amount of venous blood entering by the renal portal veins.

The lymphoid tissue which is so frequently met with in the kidneys of fishes is most abundant in the middle and anterior regions of the plaice kidney. It consists of very small cells, supported by reticular connective tissue, and filling up the interspaces between the blood vessels and the uriniferous tubules. Groups of pigment granules are scattered throughout this lymphoid tissue and give the organ its black appearance. They are small rounded granules of variable diameter, and of a greenish-black colour. They lie freely among the lymphoid cells.

The **Pronephros** and **Head Kidney** (Text-fig. 2).— The kidney in *Pleuronectes* is a mesonephros, and its paired ducts are segmental or archinephric ducts. The most common mode of origin of these structures in Teleosts is by a longitudinal evagination of somatopleure forming a groove which afterwards closes by constriction of its lips, giving rise to a tube. McIntosh and Prince state, however (*loc. cit.*), that in Gadoids and Pleuronectids the segmental ducts originate as solid rods, which afterwards acquire a lumen by the radiate arrangement of their cells. This condition, however, is most probably a secondary one, and the mode of development as a longitudinal groove seems most primitive. The cavity of the paired ducts of the adult kidney and that of the tubules is accordingly cœlomic in its nature. During larval life these ducts are the efferent channels of the pronephros— the larval excretory organ.

The Pronephros is probably formed before the larva hatches from the egg. Text-fig. 2 represents the condition of the organ in a plaice 12 days old. It is part of a transverse section through the anterior part of the trunk immediately behind the gill-bearing region. Here the segmental duct makes two or three convolutions and opens by a non-ciliated nephrostome into a small chamber. The right and left pronephric chambers lie side by side, separated by a thin septum. The dorsal aorta lies between them in the dorsal thickened part of the septum. A vascular tuft, the glomus, projects from the lateral wall of the aorta into each pronephric chamber opposite the nephrostome. The whole organ lies between the noto chord and the œsophagus. It has no connection, at least in the stage studied, with the body cavity, but there can be little doubt that the pronephric chamber is simply an enclosed portion of the general cœlom. The whole organ is essentially similar to the pronephros of *Lepidosteus* as described by Balfour and Parker, except that in the latter form the pronephric chamber still communicates with the body cavity by a richly ciliated funnel. The glomus is really a tuft of capillaries in communication with the aorta. The resemblance of the whole structure to a Malpighian body of the kidney with its contained glomerulus will be noted.

G

The pronephros degenerates at a variable age. We have seen it in a plaice of about ½ inch in length, that is at an age when the metamorphosis of the larva is complete and the adult asymmetry thoroughly established.

The Head Kidney.—In a continuous series of sections from a plaice of about one inch long passing through the whole length of the kidney, the latter organ can be seen to be divided into three ill-defined parts. The posterior or thickened portion of the kidney is crowded with uriniferous tubules cut in various planes, Malpighian corpuscles can be seen, though these are very few, and the lymphatic tissue is (relatively) not abundant. Anterior to this, and occupying the thinnest middle portion of the kidney, is a region where the segmental duct, slightly expanded, alone persists, but no uriniferous tubules are present in the sections. This is the intermediate portion of the kidney. Anterior to this, and beginning at the plane of transition of stomach into œsophagus, is a region where the segmental duct becomes thrown into convolutions. Here, too, the kidney divides into the two anterior horns which lie on either side of the œsophagus. This is the head kidney, and it contains lymphatic tissue. This tissue is present through all the length of the kidney, but is more abundant in the anterior portion, and here it is aggregated into nodules with well-marked blood chan nels between: that is, it has characters intermediate between a true lymphatic and haemolymph gland.

In the oldest specimens investigated this swollen anterior portion of the kidney has no traces of uriniferous tubules or segmental duct. It consists only of a modified form of lymphatic tissue with large blood vessels. In it are nests of black pigment in the form of irregular granules, and its wall also is deeply pigmented.

These anterior swollen portions are the degenerate

remains of the larval pronephros. This identification is confirmed by a study of various early stages. In the youngest forms examined (symmetrical larvæ one to two weeks hatched), the mesonephric tubules are absent, or are only just forming, while the segmental duct extends anteriorly as a straight line which becomes convoluted in its anterior extremity forming the pronephros. In a later stage (asymmetrical fish 3/5th inch long) the three regions of the kidney described above are well marked, the latter portion presenting all the characters of a mesonephros, while the nephrostomes and glomi of the pronephros are still recognisable, though much reduced. In an asymmetrical form about one inch in length, the intermediate lymphatic portion is relatively shorter, the mesonephric portion with its Malpighian corpuscles has extended further forward, the pronephric convolutions of the segmental duct are still present, but glomi, pronephric chambers and nephrostomes have disappeared and lymphatic tissue and blood spaces are largely developed. Finally in the oldest adult specimens examined the head kidney contains only lymphatic tissue.

The pronephros probably degenerates in all Teleostean fishes with the exception of one or two specialised forms. Organs formerly supposed to represent a persisting functional pronephros such as are present in *Lophius* have been shewn to be anterior extensions of the mesonephros, and others such as those present in *Anguilla* and *Esox* are lymphatic organs with degenerate remains of the pronephric convolutions. Only in *Fierasfer*, and perhaps *Zoarces*, is the evidence conclusive that a functional pronephros exists in adult life. In *Dactylopterus* Calderwood* has described an organ which he regards as a functional pronephros, but in this case it is still probable that

* Jour. Mar. Biol. Assoc., vol. ii. (N. S.), pp. 43-46, 1891-2.

we have to deal with an anterior extension of the mesone phros, and conclusive proof of Calderwood's hypothesis would only be afforded by tracing the embryonic pronephros into the structure so termed in the adult.

Concerning the representative of the lymphatic head kidney of Teleostean fishes in higher vertebrata, Weldon[†] put forward the suggestion that they exist in the suprarenal bodies. But he thought that the latter bodies were very generally absent among Teleosts, whereas it is now known that they are present in all forms sufficiently investigated. It is only the medullary portions which are present, however, and Grosglik[‡] has suggested that the homologues of the cortical parts of the suprarenal bodies of higher vertebrata are present in Teleosts as the lymphatic portions of the head kidney. According to Emery, this lymphatic tissue is to be derived from the peritoneal epithelium. It is a formative blastema which remains *in statu quo* on the reduction of the pronephros.

D.—THE BLOOD VASCULAR SYSTEM.

The heart in *Pleuronectes*, as in all fishes, is a respiratory one, and consists of a single auricle and ventricle. The de-oxygenated blood which it contains is propelled into the gills, and after passing through the respiratory capillary network in the branchial lamellæ reaches a great loop-shaped vessel—the circulus cephalicus—lying beneath the base of the skull. From the circulus cephalicus, which contains oxygenated blood, the carotid arteries pass forwards to supply the brain and head, the cœliaco-mesenteric artery enters the body cavity and supplies the viscera, while the rest of the body is supplied by the dorsal aorta. De-oxygenated blood,

† *Loc. cit.*   ‡ Anat. Anz., Jahrg., viii., pp. 605-611, 1885.

after traversing the systemic capillaries, returns through three main channels. The blood from the head returns directly to the sinus venosus by the jugular veins, but two portal circulations are interposed in the course of the blood returning from the viscera and the body. The caudal vein, the genital veins and other smaller vessels convey blood returning from the great muscles of the trunk and from the reproductive organs to the kidneys, where these afferent veins break up into a network of capillaries, which are in close association with the renal tubules. From the kidney the blood reaches the heart again *via* the two ductûs Cuvieri, or precaval veins; the blood from the stomach, intestine and spleen, containing the absorbed products of digestion, is conveyed to the liver by several afferent vessels known as the hepatic portal veins, and after traversing the hepatic capillaries enters the sinus venosus by the hepatic veins.

The **Pericardium and Heart**.— The pericardial cavity (*Per.* fig. 20) is displayed by dissecting away the pectoral girdles with their muscle masses, which cover it laterally and in front; behind, it is bounded by a strong fibrous septum which separates it from the body cavity. Its walls contain black pigment. The heart, which nearly fills its cavity, is suspended by the hepatic veins traversing its posterior wall, by the ductûs Cuvieri above, and by the bulbus arteriosus in front. It lies in a curved position, so that the sinus and auricle are nearly vertical, the ventricle oblique and the bulbus nearly horizontal.

The **Cuvierian Ducts** or precaval veins (*V. pc.* fig. 22) are wide thin-walled vessels passing slightly obliquely over the lateral surfaces of the œsophagus. Their union beneath the latter forms the sinus venosus (*Sin. V.* figs. 21 and 22). Sinus and precaval veins together form a horse-shoe shaped chamber surrounding the œsophagus on its lateral and ventral surfaces. In the middle of its ventral wall is the sinu-auricular orifice through which its cavity communicates with that of the auricle. This opening is guarded by a rather weak valve consisting of two membranous flaps, anterior and posterior in position, which hang down slightly into the cavity of the auricle. The large vessels opening into the sinus are the paired precaval veins into which open the paired posterior cardinal veins (*V. card.*), the paired hepatic veins and the unpaired inferior jugular vein.

The **Auricle** (*Aur.*) lies dorsal and anterior to the ventricle which it partly enfolds. Its external surface is lobulated, the postero-dorsal portion being produced into two notable lobes. Its walls are thin, but are strengthened internally, especially on their dorsal and ventral portions, by interlacing muscle bands—the musculi pectinati. A deep auriculo-ventricular groove separates it from the ventricle. Its cavity communicates with that of the latter by the auriculo-ventricular orifice, which is guarded by three semi-lunar valves—pocket-shaped membranous flaps, the cavities of which face the cavity of the ventricle. Two of these valves are large, and are nearly anterior and posterior, whilst the third is much smaller, and is situated laterally.

The **Ventricle** (*Ven.*) lies ventral and posterior to the auricle. Its walls are very thick, and are produced internally into ridges—the columnæ carneæ, which largely reduce its cavity. It is separated by a deep constriction from the bulbus arteriosus (*B.A.*), which is a flask-shaped dilatation of the proximal end of the ventral aorta. Its cavity communicates with that of the bulbus by an opening which is guarded by two strong semi-lunar valves. The wall of the bulbus is composed of fibrous connective tissue free from muscle fibres. It is very thick, and its internal surface is produced into longitudinal folds.

The **Afferent Branchial Vessels**.—The ventral aorta (*Ao. V.*) continues forward the bulbus arteriosus. It runs forward in the middle line of the body beneath the œsophagus and the ventral extremities of the gill arches. Its wall is composed of fibrous connective tissue apparently without muscle fibres. Like all the larger blood vessels in the plaice, it contains black pigment. Three afferent branchial vessels are given off at nearly equal intervals on each side. The first of these almost immediately divides into two vessels of equal calibre (*Af. Br.* 4; *Af. Br.* 3) which supply the 4th and 3rd holobranchs. Separate vessels (*Af. Br.* 2; *Af. Br.* 1) are given off to the 2nd and 1st holobranchs. The ventral aorta terminates by dividing to form the 1st afferent branchial vessels. Each afferent branchial vessel enters the gill at about one-third of the length of the latter from the ventral extremity, and immediately divides into two branches which traverse the whole length of the gill, running on the concave surface of the gill arch.

The **Structure of the Gills.**—It will be convenient to describe here the minute anatomy of the gills before considering their vascular arrangements. In the Plaice, as in most Teleostean fishes, there are four functional gills. Each gill is a holobranch, and consists of two separate series of gill filaments borne on the same branchial arch, each of which represents the demibranch or single series of filaments found on the one side of a gill pouch of an Elasmobranch fish. In the Teleostomi the septum which in the Elasmobranch separates the two adjacent demibranchs has disappeared, with the result that the two series of filaments borne by the same arch have become closely opposed.

Each branchial arch (Text-fig. 4, A.) consists of several pieces, and is for the most part a densely calcified tube, the ends of which are cartilaginous. The interior of the tube is strengthened by bony trabeculæ. The gill filaments are borne on the posterior and convex borders of the gill arches. On first inspection it may appear that there is only one series, but closer study shews that there are really two. This is particularly noticeable in the anterior gills, where the two series of filaments are of unequal length, so that all the anterior (or external) are longer than the posterior or internal ones. The bases of all the filaments borne on one branchial arch are fused together, but the greater portions of them are free from each other. In section (Text-fig. 4, B.) each filament is an isosceles triangle. They are so disposed that the apices of the triangles are directed towards each other and those of the one series alternate with those of the other. Obviously this arrangement secures the greatest economy of space consistent with the size of the filaments.

Text-figs. 4 are a diagrammatic representation of the structure of the gill filaments. Fig. A. is a diagrammatic transverse section of a gill arch, and shews two filaments belonging to adjacent demibranchs. Each filament is supported by a cartilaginous rod—the gill ray which runs down in its axis. These gill rays are superficially calcified; their proximal ends are swollen and are all fused together, but the connecting portions are not calcified. The skeleton of a demibranch is therefore a comb-like structure. The rays of the adjacent demibranchs are placed alternately, so that the knob-like calcified proximal end of one ray is placed opposite the cartilaginous connecting portion of the two opposed ones. Dense ligamentous bands connect the fused heads of the gill rays with the branchial arch and with each other.

TEXT-FIG. 4.   Structure of the Gills :

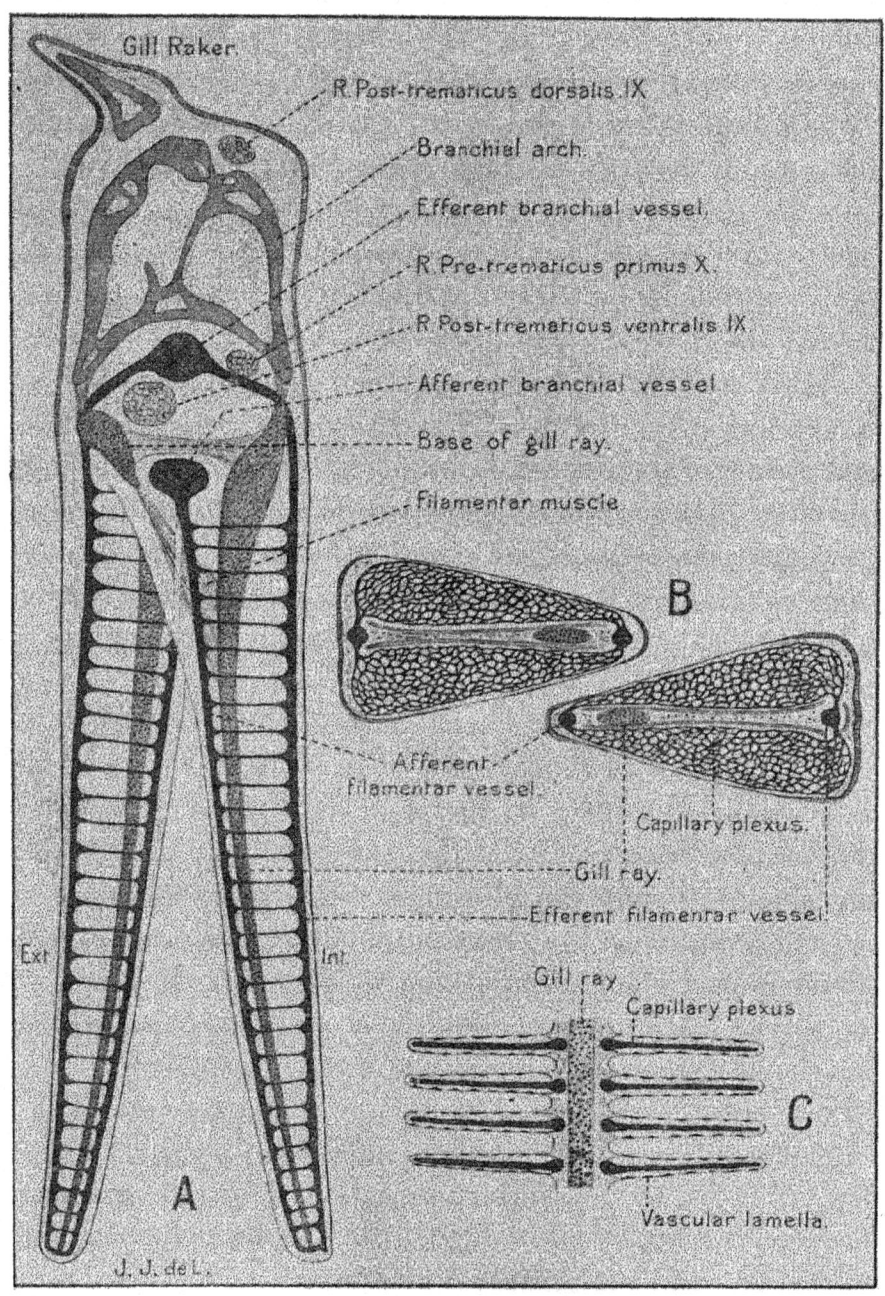

A. Diagrammatic transverse section through Branchial arch I.   B. Transverse section of a double filament, showing the surfaces of two lamellæ.
C. Longitudinal section of a single filament in a plane at right angles to A.

These structures, the convex surface of the gill arch, the proximal ends of the gill rays and the ligaments, form a tunnel through which pass the efferent branchial vessels and certain of the nerves of the gill. A very definite little muscle takes origin in the cartilaginous connecting portion of every two gill rays, passes obliquely downwards into the tissues of the opposite gill filament, and is inserted into the upper portion of its gill ray. This is part of an extremely pretty mechanical arrangement. It has been stated that each half filament is in section an isosceles triangle, and that the two series dove-tail into each other on account of their alternate arrangement. Obviously the contraction of the little muscles described above must have the effect of approximating the two series and obliterating the spaces between all the separate filaments; conversely the relaxation of the muscles and the elastic recoil of the gill rays must separate all the filaments attached to a single arch, leaving a space between each two, and this is effected without any alteration in the total length of the gill. It seems extremely probable, though we have no experimental evidence on the point, that these movements do actually take place in life, and that they aid in the movement through the gill of the respiratory water.

Text-fig. 4, C. is a longitudinal section through a gill filament in a plane at right-angles with that of 4, A. It shews that each filament consists of a flattened axis which bears on either side a close-set series of lamellæ The axis is strengthened by dense connective tissue, and contains the gill ray and the filamentar blood vessels. Text fig. 4, B. is a transverse section of a double filament between the centres of two lamellæ. It shews the position of the axis and the gill ray.

If the branchial blood vessels are injected from the ventral aorta, a series of vessels become apparent on the internal (with respect to the middle line of the gill arch) sides of the filaments. These are the afferent filamentar vessels (4, A. and B.), and they are connected with the afferent branchial vessels which run in the fused bases of the filaments outside the tunnel referred to. If, on the other hand, the system is injected from the dorsal aorta, a second series of vessels which run down on the outer surfaces of the filaments becomes visible ; these are the efferent filamentar vessels, and they are connected with the efferent branchial vessels which run in the tunnel on the convex surface of the arch. At regular intervals along its course the afferent filamentar vessel gives off an arterial twig on either side of the axis of the filament which passes into the respiratory lamellæ (4, B).

Text-fig. 4, B. represents a surface view of two lamellæ, the transverse section of the filament passing through the axis between two such lamellæ. In a fortunate injection of the branchial system it will be seen that the lamellar branches of the afferent filamentar vessel on entering the lamellæ immediately break up into very close capillary networks. This capillary network, seen from the side, is represented by the transverse black lines connecting the two filamentar vessels in 4, A. Each lamella has a wall which at the base is composed of cubical cells, but which over the flat surfaces is a thin squamous epithelium. Within the space enclosed by this wall is the capillary network, and no other tissues. According to Plehn* the blood flows in spaces hollowed out of adjacent closely-fitting cubical cells. The capillaries have not the ordinary epithelial wall characteristic of such vessels. After having traversed this network the blood is received

* Zum feineren Bau der Fischkieme. (Vorl. Mitth.) Zool., Anz No. 648 24 Bd., pp. 439-443, 1901.

by a very short vessel which opens into the efferent filamentar vessel.

Respiration is effected by rythmical swallowing movements of the mouth and co-ordinated liftings of the opercula. The water swallowed passes from the pharynx through the gill slits and over the surface of the gill filaments. Probably movements of the latter in the mode already suggested assist in this circulation of the sea water. The gaseous interchange between the blood in the respiratory lamellæ and the water takes place through the extremely thin walls of the latter and the walls of the capillaries. Only two thin epithelia separate the two liquids, and through these carbon dioxide passes from the blood to the sea water, and oxygen from the sea water to the blood.

The respiratory area of the gills can be approximately calculated. The lengths of the gill filaments vary, and the greatest number of lamellæ counted on any one side of a filament was 225; probably 150 will represent the average number to a side of a filament; there are two series of lamellæ, of course, on each filament. Therefore we have:—

Holobranchs.	No. of Filaments.	No. of Lamellæ.
I.	83 × 4	99600
II.	78 × 4	93600
III.	72 × 4	86400
IV.	58 × 4	69600
Total ..	1164	349200

Now the area of each lamella can be approximately calculated, since it is nearly triangular. It is roughly 0˙365 square millimetre. But since both the flat surfaces of the lamella are in contact with the water, the respiratory surface is double this, and is 0˙730 sq. mm. × 349,200 = 254,916 sq. mm. That is over $\frac{1}{4}$ square metre. The total respiratory surface of the gills is therefore that of a square, the length of the side of which is $\frac{1}{2}$ metre. These calculations apply to a plaice of about 22 inches long. The area of the skin of such a fish is approximately 2,340 sq. cm., or nearly $\frac{1}{4}$ sq. metre. The respiratory surface of the gills is therefore about equal to the total area of the skin.

The **Efferent Branchial Vessels.**—The blood, after having passed from the heart and afferent vessels through the lamellar capillaries, is collected by four trunks on each side—the efferent branchial vessels. These open into the epibranchial arteries of each side. Posteriorly the two epibranchial arteries (*A. ep.*) unite to form the dorsal aorta; anteriorly they are connected together by a short anastomosing vessel (*Cir. c.*). The loop thus formed is the circulus cephalicus. It is the reservoir into which the blood, after having undergone oxygenation in the gills, is poured, and from which it is distributed over the body. The efferent branchial system is best injected from the dorsal aorta after tying the cœliaco-mesenteric artery. It can be displayed after cutting away the greater portion of the operculum of one side, removing the opercular, sub-opercular and inter-opercular bones. The remaining dorsal portion of the operculum is then forced outwards and held in position by a hook. The gill filaments should be cut away close to the arches. The vessels themselves are then seen, after dissecting apart and removing most of the muscles, passing dorsally from the gill arches. The circulus cephalicus and carotid arteries can be dissected by removing the greater portions of both opercula as indicated above, and then cutting away the gills, having previously divided the efferent vessels as far away from their attachments to the epibranchial arteries as possible. The head is then placed ventral side uppermost, and held in that position by hooks. The ventral portion of the parasphenoid must be removed in order to study the course of the internal carotid trunks.

The first and 2nd efferent branchial vessels (*Ef. Br.* 1, *Ef. Br.* 2) open separately into the epibranchial trunk. The 3rd and 4th (*Ef. Br.* 3, *Ef. Br.* 4) unite together to form a short trunk. On the left side this opens into the left epibranchial, on the right it opens into the cœliaco-mesenteric artery (*A. cm.*). It may appear, however, that the cœliaco-mesenteric, instead of taking origin from the epibranchial as represented in the figure, springs from the common trunk of 3rd and 4th efferent branchials.

Immediately behind the union of the epibranchial trunks the subclavian arteries are given off from the dorsal aorta. Each of these vessels (*A. scl.*) passes out transversely, then bends down ventrally and runs along the internal surface of the corresponding pectoral girdle, the muscles of which it supplies. Behind these vessels transverse arteries are given off from the dorsal aorta on either side, one to each segment. These vessels are not represented in the figure.

Several arteries leave the epibranchials to supply the muscles of the gill arches with blood. The most important of these is a paired vessel which leaves the epibranchial immediately anterior to the place of entrance of the 2nd efferent vessel. It passes at first dorsally, then backwards and downwards over the 3rd and beneath the 4th efferent trunks. Approaching the middle line it runs ventrally among the muscles on the anterior border of the pericardium, where it apparently breaks up; a much smaller vessel takes origin from the dorsal portion of the 1st efferent vessel and runs backwards and downwards among the muscles of the 1st and 2nd gill arches, where it breaks up. These vessels are represented but not lettered in fig. 22.

Two fairly large trunks take origin on each side from the ventral portions of the 1st and 2nd efferent vessels. Apparently they do not anastomose in *Pleuronectes*. The first, which is the hyoidean artery (*A. hy.*), leaves the efferent trunk while still within the arch, and after giving off a small twig, which breaks up on the internal surface of the operculum, turns round dorsally and runs on the internal surface of the operculum externally to the cerato-hyal and symplectic bones. At the level of the upper extremities of the gill arches it breaks up into a number of branches which end in the filaments of the pseudo-branch (*Ps. Br.*).

The **Afferent Pseudobranchial Vessel.**—The precise disposition of the afferent pseudobranchial vessels varies among Teleostean fishes. In the greater number the afferent vessel is the hyoidean artery, which, moreover, anastomoses with the circulus cephalicus, so that the blood in the minute vessels of the pseudobranch may be derived from that in all the efferent branchial vessels. This is the arrangement in *Gadus*. In others, of which *Salmo* is an example, the hyoidean artery is the sole afferent vessel, and does not anastomose with the circulus cephalicus. In addition to these types of blood supply, Maurer* has described another in *Esox*, where the afferent vessel of the pseudobranch is a twig of the circulus cephalicus and the

* Beitr. zur Kenntniss der Pseudobranchien der Knochenfische, Morph. Jahrb., 9 Bd., pp. 229-252, 1883-4.

hyoidean artery contributes nothing. In *Pleuronectes* the organ is supplied by the hyoidean artery, and there is no evidence of an anastomosis of the latter with the circulus cephalicus beyond a doubtful communication between branches of the hyoidean and external carotid arteries.

The **Efferent Pseudobranchial Vessel** is the ophthalmic artery. Blood, after traversing the capillaries in the pseudobranchial filaments, passes into this vessel (*A.op.*), which runs forwards along the roof of the pharynx covered over by the mucous membrane. The two ophthalmic arteries approach each other in the middle line of the body, then separate and perforate the prootics together with the superior jugular veins, passing through the jugular foramina (*f. jug.* fig. 2). Each vessel accompanies the jugular vein and optic nerve of its side, and reaching the eye perforates the sclerotic and breaks up in the choroid gland. This peculiar arrangement is common to all Teleostean fishes, and has not received any satisfactory explanation. Joh. Müller suggested that the pseudobranch was a gland furnishing an internal secretion, and that the object of the included capillary system of the pseudobranch was to equalise the intra-optical pressure by smoothing down the pulsations of the heart. But the blood in the ophthalmic artery has already passed through the branchial capillaries before reaching the pseudobranch, and there is no evidence of the elaboration of any internal secretion.

The **Pseudobranch.**—It will be convenient to give some account here of the structure of this organ. It is situated on the inner surface of each operculum in a little concavity which lies behind the strong transverse muscles forming the roof of the pharynx, and which is formed by the abrupt termination of these in a posterior transverse ridge. It is situated mostly on the hyomandibular, but its ventral extremity lies on the preoperculum. It is so situated that its attached base is exactly opposite to the dorsal portion of the first branchial arch, and the direction of its filaments is almost exactly that of those of the dorsal portion of the first holobranch, that is, posterior and slightly dorsal. The first branchial cleft is therefore bounded posteriorly by the anterior demibranch of the 1st branchial arch and anteriorly by the pseudobranch. The afferent pseudobranchial vessel or hyoidean artery runs along the external or deep-seated part of the base of the pseudobranch, and gives off a vessel to each filament. The efferent pseudobranchial vessel or ophthalmic artery runs along the internal or visible part of the base, and receives a vessel from each filament.

The filaments of which the pseudobranch is composed are strikingly similar in appearance to those of any one of the true demibranchs, and their structure is the same in all essential points. Each is made up of a flattened axis, on each side of which are borne a number of lamellæ. The afferent filamentar vessel runs down the internal (with respect to the attachment of the organ to its arch) edge of this axis; the efferent vessel runs up the external edge. Small twigs are given off from the afferent vessel into each lamella, and in each of the latter they break up into a capillary plexus, as in the true gills, which empties its blood into a corresponding twig opening into the efferent filamentar vessel.

Only about one-half of each filament projects freely into the opercular cavity. The basal halves are all attached to each other and to the epithelium clothing the inner surface of the operculum. The lamellæ are mostly free, but many are attached together by their edges. They differ from the lamellæ of the true gills in that their wall instead of being a squamous epithelium is made

H

up of nearly cubical cells and contains numbers of goblet cells, in this respect resembling the general wall of the pharyngeal cavity or the epithelium covering the branchial arches.

Probably the pseudobranch is not a functional respiratory organ, though its structure is very similar to that of any demibranch of the posterior series of true gills. The walls of the vascular lamellæ resemble mucous epithelia rather than membranes through which gaseous interchange may take place. And the blood reaching the organ has already passed through respiratory plexuses in the first holobranch. Undoubtedly it is part of the holobranch which was formerly situated on the hyomandibular arch, and its situation suggests that it is the posterior demibranch of that gill. There are no traces of the presence of a vestige of the anterior demibranch of this gill, and the structure of the organ is exactly that of a demibranch, the respiratory surfaces of which are greatly modified. Since no traces of the afferent vessel originating in the ventral aorta, which would have supplied a functional hyomandibular gill, are present, the vascular supply gives no certain indication of the homology of the organ.*

The **Pelvic Artery**.—Each of the 2nd efferent branchial vessels gives origin to an artery which almost immediately unites with its fellow of the opposite side, and the azygos trunk so formed runs backwards in the floor of the pharynx in the middle line of the body. Various small vessels are given off to the ventral portions of the branchial arches. This pelvic artery (*A. pe.*) then gives origin to a small vessel supplying the pericardium, the pericardial artery (*A. per.*), and continues backwards between the

---

* Cp. the cranial nerves for a discussion of the nervo supply of the pseudobranch.

pelvic arches almost to the musculature of the body wall surrounding the anus.

The **Carotid Arteries**.—The portion of each epi-branchial artery anterior to the entrance of the 1st efferent branchial vessel may be spoken of as the common carotid artery. It is a very short trunk which divides into two vessels. The outer of these, the external carotid (*A. Car.*[1]), curves round behind the pharyngo-branchial segment of the 1st branchial arch, and runs forward on the ventral surface of the skull. Several branches are given off which break up on the internal surface of the operculum and on the base of the skull. The internal branches of the common carotids, the internal carotid arteries (*A. car.*), after perforating the skull at the junction of the prootics and parasphenoid by the carotid foramina (*f. car.* fig. 2), communicate by a very short anastomosing vessel (*Cir. c.*) which completes the circulus cephalicus. From this transverse anastomosing vessel three arteries take origin, which run anteriorly in the trough of the parasphenoid. The two external vessels, which are the internal carotid arteries, run forwards towards the nasal region of the skull. The internal median vessel divides, and the two vessels so formed run forwards in the trough of the parasphenoid, or eye muscle canal, accompanying the eye muscles. Each passes out with the corresponding optic nerve, and runs forwards towards the eye.

The **Visceral Arteries**.—The cœliaco-mesenteric artery (*A. cm.*) is an unpaired vessel lying entirely to the right side of the body. After leaving the right epibranchial artery it passes over the external surface of the right precaval vein, and gives off a small branch—the œsophageal artery (*A. œ.*), which breaks up on the wall of the œsophagus. It then almost immediately bifurcates, and from the point of division the cœliac artery (*A.cœ.*) is given off. From this artery a small vessel arises (*A.hep.*) which turns backwards and enters the liver on the posterior surface of that organ. The two cœliaco-mesenteric trunks then pass internally to the right lobe of the liver. One vessel runs in the mesentery, giving origin to branches which supply the greater portion of the intestine from the anus forwards. The other courses in the mesenteric sheet connecting the liver with the duodenal loop, and supplies that portion of the intestine and the stomach.

The **dorsal aorta** gives origin to a pair of arterial trunks in each segment, which supply the muscles of the trunk. Towards the posterior extremity of the kidney, a large median vessel—the common genital artery (*A. gen.*) —is given off, and passes downwards through the posterior portion of the kidney, sending small branches to the kidney and suprarenal bodies. This divides into two branches, one of which goes to each ovary or testis and the adjacent portions of the body wall. The dorsal aorta then passes backwards to the tail in the tunnel formed by the haemal arches.

With regard to the venous system, we propose to describe the larger venous trunks only. All the blood from the head is returned to the heart *via* the paired superior and the unpaired inferior jugular veins.

The **Superior Jugular Veins** (*V. Jug.*) are large thin walled vessels which will have been exposed in dissecting for the branchial vessels. They receive the blood from the eyes and adjacent parts, and accompany the eye muscles in the eye muscle canal, emerging from the latter through the jugular foramina (*f. jug.* fig. 2). They then run backwards on the ventral surface of the skull over the dorsal extremities of the branchial vessels slightly dorsal and external to the epibranchial arteries, receiving in their course vessels conveying blood from the head and brain, and enter the precaval veins at the dorsal and anterior extremities of the latter.

The **Inferior Jugular Vein** (*V. Jug.*[1]) is an azygos trunk running backwards under the ventral wall of the pharynx immediately above the dorsal aorta. It then passes upwards on the anterior wall of the pericardium, and may enter either the right or left side of the sinus venosus, though its ending on the right side seems to be the more common one.

The **Hepatic Veins** (*V. hep.*) are short wide trunks coming from the liver, which penetrate the posterior wall of the pericardium and enter the posterior part of the sinus venosus.

The **Renal Portal System.**— The afferent vessels of this system are the parietal veins, the caudal vein and the genital veins. The caudal vein (*V. cd.*) runs forwards from the tail in the haemal canal immediately beneath the dorsal aorta. It enters the kidney at the most dorsal and posterior angle of the latter organ, and divides into two vessels which run forwards in the kidney and break up, but do not apparently anastomose with the cardinal vein. A short venous trunk comes from each ovary (or testis) and the adjacent portions of the body wall, and enters the kidney on each side near the extreme ventral tip of the latter organ. A series of veins from the muscles of the trunk enter the dorsal portion of the kidney on each side; these are the parietal veins. One such vessel is represented in fig. 22 as entering the anterior tip of the head kidney.

The efferent vessels of the system are the cardinal veins, which run forward in the kidney. The right cardinal vein (*V. card.*) runs along the middle part of the kidney, and is visible on its ventral surface. We have said that the kidney at its anterior extremity is divided into two horns, which reach forward towards the heart. The right cardinal vein emerges from the kidney through the right horn and enters the posterior side of the right precaval vein. The left cardinal (as in the Cod) is a short vessel which begins at about the anterior third of the kidney, traverses the left horn, and enters the posterior side of the left precaval vein. The two cardinals do not apparently anastomose with each other.

The Hepatic Portal System.—The afferent vessels of this system are the portal veins carrying the blood from the stomach, intestine and spleen. The smaller factors of this system have much the same course and distribution as the branches of the cœliac and cœliaco-mesenteric arteries. They do not, however, unite to form a single hepatic portal vein, but enter the liver as a variable number of separate portal veins. Commonly there are (1) a trunk receiving the blood from the spleen and the greater portion of the intestine, and anastomosing with (2) a vein receiving the blood returned from the loops of the intestine posterior to the pylorus; (3) a smaller vessel draining the region of the pylorus, and (4) a vein coming from the stomach. These vessels enter the internal surface of the liver principally on the larger left lobe, and run for some distance parallel to and immediately beneath the surface, so that their ramifications can be easily traced. Their precise number and distribution in the liver varies; five such trunks are represented in fig. 21, cut off close to the liver surface ($Vp.$) The apparent calibre of the intestinal veins, and to a less extent the arteries also, is increased by the presence of the perivascular glandular tissue referred to above. The efferent vessels of the hepatic portal system are the two large paired hepatic veins (*V. hep.*) which enter the lower portion of the sinus venosus on its posterior side.

E.—THE NERVOUS SYSTEM.

We shall commence our description of the nervous system with the brain and spinal cord, then proceeding to the cranial and spinal nerves, and finally to the sympathetic nervous system.

1.—THE BRAIN AND SPINAL CORD.[*]
(Figs. 28, 30, 31).

The brain of the Plaice may be conventionally divided into four regions, including the following structures:—

A. **Hind - Brain.**—This comprises the medulla oblongata, which itself includes many structures that can only be regarded as the continuations of corresponding ones in the spinal cord, and the cerebellum. The latter consists of a body and the anterior valvula cerebelli.

B. **Mid - Brain.** —Formed by a base (crura cerebri) and side wall, and the tectum opticum or tectum mesencephali (optic lobes).

C. **'Tween-Brain.**—Represented by three parts: (1) the epithalamus (epiphysis generally and the ganglia habenulæ); (2) the thalamus (optic thalami—thalamencephalou); (3) the hypothalamus .(corpus geniculatum,

---

[*] The following works will be found to contain references either to the brain of the Plaice or to allied Pleuronectids :—Cattie, Arch. Biol., iii., p. 150; le Roux, "Recherch. Syst. Nerveux Téleostécns," Caen, 1887; Mayne, "Optic Nerves," Todd's Cyclopædia, part xxvi.; Malme, Bibang K. Svens. vet.-akad Handlingar, xvii.; Mayer, Verhand. K. Leop.-Carol., xxx.; and Steiner, "Entstehung d. asymmetrischen Baues der Pleuronectiden," 1886; a recent important work, by J. B. Johnston, on the brain of *Acipenser* (Zool. Jahrb., Abth. Morph., xv.), may be used as a starting point in studying the brain of Fishes in detail.

corpus mammillare, infundibulum, lobi inferiores, saccus vasculosus and pituitary body).

D. **Fore-Brain.**—This may be considered as including the epistriatum, striatum proper and the membranous pallium, together with the bulbus olfactorius.

The roots of the cranial nerves will be described in the section on the nerves.

In a dorsal view of a well-preserved brain we note the following characters:—First the relatively small size of the brain. This is seen also in the Cod and in Teleosts generally. The small brain lies in the large cerebral cavity, surrounded by a packing of areolar connective tissue loaded with fat, and seems to be very disproportionate to the size of the fish. Then the asymmetry of it is at once striking. The spinal cord, on entering the brain case, turns slightly to the left, but opposite the cerebellum it swerves markedly to the right, so that a median line would pass through the left striatum instead of between the two striata.

In the medulla the great reduction of the terminal bud system that has taken place involves the absence of the lobi vagi. Also the lateral line system is not sufficiently robust to have produced that exaggeration of the tuberculum acusticum known as the lobus lineæ lateralis. The medulla is therefore smooth, and presents no obvious traces of its ganglia. On removing the vascular covering of the fourth ventricle known as the choroid roof, the ventricle itself is seen to be apparently divided into two parts by the partial union over its roof of the medio-lateral portions of the tuberculum acusticum, forming an elliptic-shaped opening behind (calamus scriptorius) and a triangular one in front, with its apex directed backwards. The cerebellum, of which the body only is visible in the undissected brain, is small and globular. This is what one would expect, seeing that it is connected with the general activity of the organism, and the Plaice is sluggish in habits.

The mid-brain is represented on the dorsal surface on each side by the tectum opticum (optic lobe). These are very large bodies almost spherical in shape, and characterised in dead and preserved specimens, and doubtless in life also, by a deep furrow, which extends backwards in a curve from the anterior margin of each lobe for about half its antero-posterior diameter.

As is usual in Teleosts, the 'tween-brain hardly appears at all on the dorsal surface of the brain, being excluded from it by the meeting of the two striata and optic lobes. However, a small portion of its membranous roof is visible, and from this there is seen emerging by the triangular space formed immediately in front of the median apposition of the two optic lobes, the extremely fine pineal tube. In sections it is seen to arise as an evagination of the roof of the third ventricle almost behind the ganglia habenulæ and in front of the posterior commissure. It then passes forwards over the pallium of the left striatum and swells into the large pineal gland lying on the pallium near the anterior extremity of the left striatum. By pressing apart the optic lobes there may be seen immediately in front of the exit of the pineal tube the ganglia habenulæ and the plaited choroid roof of the third ventricle.

In a well-preserved brain the membranous pallium of the fore-brain is very obvious. It is a large oval sheet, with its long axis at right-angles to that of the brain, and almost equal to that of the optic lobes. It is a very thin membrane, and appears thicker than it really is on account of the coagulated cerebro-spinal fluid in the ventricle. The corpora striata are also visible through the pallium.

On removing the pallium, it will be noted that there are no lateral ventricles, but a large single median prosocoele. In the floor of this are raised up the large corpora striata separated dorsally by a wide fissure, but connected below by the anterior commissure, and constituting the solid cerebral hemispheres of older authors. These are considerably smaller than the optic lobes, and the dorsal surface of each is marked by a somewhat complex furrow ("sulcus" of former authors—see fig. 30). In front of and below the striata are the olfactory bulbs, from which the olfactory nerves originate. The left is smaller than the right.

On the ventral surface of the brain the most noticeable structures are the appendages of the 'tween-brain. The lobi inferiores are a pair of large bean-shaped bodies opposed by their median surfaces. In the middle line immediately in front of these is the spherical pituitary body. The apposition of the pituitary body and lobi inferiores is not complete, and a triangular space is left by which there emerges on to the ventral surface of the brain the red thin-walled saccus vasculosus. This is at its origin a very narrow tube, but it expands and passes straight backwards in the middle line over the opposed lobi inferiores. It is dilated behind, and ends blindly slightly posterior to the hinder border of the lobi inferiores. In the adult the pituitary body (hypophysis cerebri) and saccus vasculosus are essentially glandular organs receiving a marked nervous supply from the infundibulum. According to most recent authors the saccus at least "probably forms part of a mechanism for secreting, or otherwise controlling the pressure of, the cerebro-spinal fluid. It may affect the heart beat and blood pressure by way of the vagus" (J. B. Johnston).

The crossing of the optic nerves is very obvious in the

Plaice, as it is effected some distance in front of the pituitary body, and is not hidden by the olfactory nerves on the ventral surface. It is also quite clear that they merely cross and do not exchange fibres, whilst their plaited nature is at once revealed by a little simple dissection. On removing the optic nerves the two small and asymmetrical-olfactory bulbs are well seen lying largely under the anterior extremities of the two striata. In the medulla the ventral fissure of the spinal cord is continued as far forwards as the base of the lobi inferiores, where it slightly expands.

Regarding the ventricles of the brain, the central canal of the spinal cord appears in the sections as a pin hole. It begins to widen rapidly into the fourth ventricle (myelocoele) at about the posterior region of the auditory organ. The ventricle is at first very deep from above downwards and very narrow from side to side. It soon opens above, and is only closed in by the choroid roof. The peculiarity of the roof of this ventricle has been already mentioned. In front of the expanded portion it becomes completely roofed over by the tuberculum acusticum, and at the same time is reduced to a very small size. Opposite the junction of the medulla and cerebellum it again expands, but does not communicate with a cerebellar cavity (metacoele), the cerebellum being solid. In front of the body of the cerebellum it passes into the aqueductus Sylvii (mesocoele—iter a tertio ad quartum ventriculum), roofed over behind by the valvula cerebelli and communicating on each side and in front with the large space enclosed by the tectum opticum (optocoele). In front, the latter opens below into the third ventricle (thalamocoele), bounded laterally and below by the thalamus (optic thalami) and above in front by the choroid roof. The third ventricle is prolonged downwards and somewhat backwards into the hollow infundibulum. If this be now traced posteriorly, it is found first of all to communicate with the cavity of the pituitary body. Almost at the same time, but more dorsally, it is prolonged on each side into the large cavities of the lobi inferiores, whilst finally it communicates with the cavity of the fine stalk of the saccus vasculosus. The third ventricle therefore is continuous with the cavities of the pituitary body, lobi inferiores and saccus vasculosus. The infundibulum of the Plaice is difficult to delimit, as it is largely merged into the floor of the thalamus. Incidentally we may draw attention in the latter to the very large paired nucleus rotundus, which is very striking in sections. Anteriorly the third ventricle passes into the large median ventricle of the fore-brain (prosocoele), roofed over by the pallium. The prosocoele is not prolonged into the bulbi olfactorii as a rhinocoele, the bulbs being solid.

Cunningham makes two assertions on the brain of the Sole that appear to us to require confirmation. One is that the "position of the brain is almost entirely unaffected by the change which has taken place in the normal position of the fish," and the other is that "the left olfactory lobe is somewhat larger than the right, a difference which is related to the great development of the left olfactory capsule." On the other hand, Malme states of the Sole (*op. cit.*, p. 34) that "insbesondere . . . . . ist der rechte Lobus [striatum] (derjenige der Augenseite) viel grösser als der linke," and again that in Pleuronectids generally "der Bulbus der Augenseite ist stets der grösste." Malme's observations agree with ours on the Plaice. Again, Cunningham apparently overlooks the work of Rabl-Rückhard on the brain of Teleosts, and describes what are really the corpora striata as receiving prolongations from the third ventricle.

In the **Spinal Cord** we wish to direct attention to two peculiarities only. The first is the giant ganglion cells that are found in the dorsal fissure. Transverse sections of the cord will demonstrate these quite easily. They have been studied especially by J. B. Johnston,[*] Sargent[†] and Dahlgren.[‡] The latter author, who has devoted his attention particularly to the Pleuronectidæ, states that these very peculiar cells are the first ganglion cells to be differentiated in the embryo flat-fish, and that they become an important and permanent apparatus in the adult. In an adult fish they are seen to form a row of very large nerve cells in the median dorsal fissure, and their neurites pass backwards to form an isolated fibre tract on the median side of each dorsal horn. Their exact distribution and function are unknown, but Dahlgren suggests that the neurites pass out with the dorsal roots of the spinal nerves and are connected with the sensory supply of the unpaired fins. Sargent finds in *Ctenolabrus* that the giant cells are connected with a fibre bundle passing forwards through the cord and medulla, and emerging by the ventral root of the trigeminus nerve. If this be true then the fifth nerve of this fish possesses a nerve component not hitherto recognised, and it would be interesting to study the giant cell apparatus from the point of view of the component theory.

The second peculiarity of the cord is one which it shares with all Teleosts, and that is in the presence of the very interesting rod or fibre within the lumen of the central canal known as Reissner's fibre. This fibre has been investigated recently by Sargent,[§] who finds that it " extends through the whole length of the canalis centralis

[*] Jour. Comp. Neurol., x., p. 375. [†] Anat. Anz., xv., p. 212.
[‡] Anat. Anz., xiii., p. 281.
[§] Anat. Anz., xvii., p. 33, and Proc. American Acad. Arts and Science, xxxvi., No. 25,

of the cord and continues cephalad through the 4th and 3rd ventricles to the anterior end of the optic lobes," where it passes into the brain tissues. It is not a single fibre, but a collection of axis cylinders, and is therefore a fibre tract. Some of the fibres in the tract originate in cells situated at the posterior extremity of the central canal, and pass forwards to the tectum opticum. Others originate in cells in the tectum opticum and pass backwards as far as the "posterior canal cells." The tract, therefore, contains fibres coursing in two opposite directions. According to Sargent this unique apparatus forms a "short circuit between the visual organs and the museu lature, and has for its function the transmission of motor reflexes arising from optical stimuli." It is most highly developed in active fish, and is entirely absent in the blind vertebrates of the cave fauna.

2.—The Cranial Nerves (Fig. 23).

In spite of the fact that the cranial nerves of Fishes have been more or less investigated for about two and a half centuries, it is only within the last few years that our knowledge of them has assumed a form likely to be at all lasting. Although these results were made possible as long ago as 1811 by the enunciation of Bell's law, and although this law was very ingeniously developed and applied to Fishes in 1849 by Stannius, who has never received due credit for his work, it was only in the eighties that Gaskell stated his "four root theory" of the spinal nerves, which showed that there were represented in each spinal nerve four kinds of fibres instead of the two assumed by Bell's law.

The attempt to strictly apply the four root theory to the cranial nerves of lower vertebrates has not only been unsuccessful, but it has actually retarded knowledge by diverting the energies of investigators into an unprofitable channel. The work on the cranial nerves of the frog's tadpole, published in 1895 by Strong, distinctly proved this, for he showed that, for example, there were three systems of sensory fibres in the cranial nerves of the larval frog, one of which must be considered characteristic of the head and not represented in the spinal nerves at all, and another only partly so.

One of the first results of Strong's work was to show that the old system of classifying the cranial nerves of Fishes into ten formal pairs was essentially unsatisfactory, and that attention should be concentrated rather on the various definite systems of nerve fibres characterised by their structure, central origin and peripheral distribution, than on those heterogeneous collections of nerve rami known as the "cranial nerves." We must, however, in the meantime adhere to the old classification, until sufficient work has been carried out on the new lines to justify a revision of the cranial nerves, and to ensure for its findings some permanent value.

The new theory of the cranial nerves is known as the "component theory." It takes advantage of the fact that the fibres forming them, and omitting the olfactory and optic nerves and the sympathetic, which present problems of an altogether special nature, fall by reason of their functions and certain structural relations into five fibre systems, three of which are sensory and two motor. Each system is delimited by a uniformity of peripheral termination and a special and characteristic origin in the brain, and each system may appear in a variable number of cranial nerves as a component of those nerves. It is therefore indispensable, as we have done in the Plaice, to work out the whole course of the nerves by means of serial sections. Microscopic work either on the brain or the peripheral nerves only is inadequate, and dissection, as a means of research, has but a very doubtful value. The only fish which has been thoroughly investigated according to the component theory is *Menidia*—in an important work published recently by C. J. Herrick, who truly remarks: "Until each component can be isolated and treated as a morphological unit, and then unravelled in its peripheral courses through the various nerve roots and rami—until this is possible, no further great advances in cranial nerve morphology can be looked for even among the lower vertebrates, still less in man."

The five systems of fibres which variously compose the cranial nerves of the Plaice are as follows:—

1. **General Cutaneous or Somatic Afferent System.**— These fibres, which undoubtedly correspond to the cutaneous fibres of the spinal nerves, are derived from continuations of the dorsal horns of the spinal cord, which form two longitudinal bundles in the medulla known as the spinal vth tracts. These fibres in the Plaice leave the brain by the roots of two cranial nerves only—the vth and the xth. In the former case their ganglion is the Gasserian ganglion, in the latter the jugular ganglion. The cutaneous fibres in the facial nerve are distinctly derived from those of the fifth. The fibres of this system are distributed generally to the skin, and do *not* end in any specialised dermal sense organs. Hypertrophy of this system produces a corresponding hypertrophy of its centre in the central nervous system, as witness the remarkable lobes at the anterior extremity of the spinal cord of *Prionotus* (Morrill).

2. **Somatic Efferent System.**—Represented by the heavily myelinated eye muscle nerves (iii., iv. and vi.). This system is of course largely present in the so-called

" hypoglossal " nerve, or first spinal, but we do not consider this to be a cranial nerve in fishes.

3. **Communis (Viscero Afferent?) System.** — Partly synonymous with the fasciculus communis system of Osborn and Strong. A striking feature about this sensory system is that it may innervate both ecto- and endo-dermal surfaces, and it may therefore be disputed whether it is a visceral nerve that has invaded the skin, a somatic nerve that has invaded the visceral surfaces, or a complex of more than one component. The latter seems perhaps the most probable. The fibres of the communis system are fine and lightly myelinated, and are distributed peripherally as follows :—(a) to the special sense organs in the outer skin called " terminal buds," *i.e.*, to all the definite sense organs of the skin not belonging to the lateral line system. This part of the component has been reduced in the plaice ; (b) to taste buds in the mouth ; (c) to the general mucous surfaces without the intervention of sense organs at all. The ganglia and cranial nerves into which the system enters are : (a) the genienlate ganglion (vii.), glossopharyngeal ganglion (ix.), and the intestinal and four branchial ganglia of the vagus (x.). Any communis fibres in the trigeminus arise from the communis facialis. The central origin of the component is the Lobus vagi, and the enormous vagal lobes of *Carpiodes* are simply due to the hypertrophy of the communis vagi component in this fish (Herrick). Further the so-called Lobus trigemini of some fishes (*Amiurus*) is due to the hypertrophy of the communis facialis, and hence it should be called Lobus facialis.

4. **Viscero Efferent System.**—This comprises the motor roots of the vth, viith, ixth and xth cranial nerves. Each of the first two has its own motor nucleus in the brain, but the two latter arise from collections of cells

supposed to represent the nucleus ambiguus. The fibres of this system are heavily myelinated.

5. **Acustico Lateral System.**—Includes the auditory and lateral line nerves. Let it be emphasised at once, what is taking a long time to filter down into the text-books, that the lateral line nerves can only be associated with two cranial nerves—the viith and xth. The lateral line fibres in the fifth nerve are always derived from the facial, those in the ixth (when present) from the vagus. The fibres of this system are distributed to the ear, to the sense organs in the lateral line canals, and to those lateral line sense organs lying free on the skin, and known as pit-organs. Its ganglia are the dorsal and ventral lateral ganglia of the facial, and the lateralis ganglion of the vagus, and its fibres are very large, being in fact the coarsest in the body. Its central termination is the tuberculum acusticum of the antero-dorsal region of the medulla—associated with the cerebellum. The hypertrophy of the lateral line nerves produces an exaggeration of the tuberculum acusticum well marked on the surface of the brain and called by Johnston the Lobus lineæ lateralis. This lobe has also been called the Lobus trigemini by the older writers, and when associated with lateral line fibres it may well receive the name given it by Johnston. Otherwise it should be called the Lobus facialis (see above).

Nervus Olfactorius*—I. (Figs. 25 and 28.)

As considerable asymmetry is exhibited by these nerves, both sides will be described.

*For the cranial nerves of Teleostean Fishes compare especially the following works;—Allis (*Amia*), Jour. Morph., ii. and xii. ; Cole (*Gadus*), Trans. Linn. Soc., ser. 2, vii. ; Desmoulins and Magendie (nerves of *Rhombus*), Anat. Syst Nerveux, Paris, 1825 ; Herrick (*Menidia, Gadus, and Amiurus*), Jour. Comp. Neurol., ix., x., and xi. ; Juge (*Silurus*), Rev. Suisse Zool., vi ; and Stannius (general), Rostock, 1849. The works of *Herrick* are most important so far as the Plaice is concerned, and should certainly be consulted. We have purposely adopted, as far as possible, the same reference letters, in order that the comparisons may be facilitated.

The right Bulbus olfactorius† lies mostly under the corpus striatum (the latter is the cerebral hemisphere of older authors). Behind, it is free, unconnected with the striatum and ends bluntly, but in front it acquires a firm connection with the striatum. Anterior to this again it separates once more from the striatum. So far it has been increasing in size, but it now begins to taper down, its ventral portion becomes fibrous, and its dorsal divided into two. The upper or cerebral portion disappears in front, and the remainder narrows down into the cylindrical nervus olfactorius. Both olfactory nerves lie to the right of the upper or left optic nerve. As the nerve passes forwards it becomes divided by connective tissue strands into two or more fasciculi, each of these again being further subdivided into small bundles of fibres. The right olfactory passes through the foramen olfactorium in the right prefrontal, turns up at once and breaks up in the olfactory laminæ of the right nasal chamber.

The left Bulbus olfactorius is not free behind like the right, but passes imperceptibly into its striatum. Nor is it situated below the latter, but between the two striata (see fig. 28). The appearance therefore of this portion of the brain is very asymmetrical, and suggests a rotation towards the right side of the ventral axis of the brain only. The left bulbus is perceptibly smaller than the right, but the left striatum extends further forwards than its fellow. The bulbus separates from the striatum in front, becomes fibrous at its right ventral corner and gives off the left olfactory nerve, which passes at once to the right side, so as to lie near the right bulbus. The left

---

† This structure is also called by some authors the *Tuberculum olfactorium* (Stannius) and *Lobus olfactorius*. We have no space to discuss the precise significance of each of these three terms, if indeed they have any (but see Elliot Smith, Jour. Anat. and Phys., xxxv., 1901).

bulbus has no direct connection with its nerve in front as on the right side, nor does it extend as far forwards as the right one. There is much more difference in the size of the two olfactory nerves than one would expect from the sizes of their bulbs; the left being only $\frac{1}{4}$ the size of the right. Nor do its fibres take such an intense stain with the osmic acid. For some distance the two olfactory nerves course together, but finally the left separates from the right and passes upwards towards the eyeless side of the body. It then traverses the olfactory foramen in the left prefrontal, and at once passes straight upwards to break up in the olfactory laminæ of the left nasal chamber. The left nasal organ is much smaller than the right (cp. fig. 25, *n. olf.*, *n. olf.*[1]), and hence the small left nerve. It is also situated somewhat behind the right, and therefore the left olfactory is the shorter of the two.

Nervus Options—II.

As in all lower vertebrates, the fibres of the optic nerve arise mostly from the roof of the mid-brain (tectum opticum), and as is usual in Teleosts they pass forwards over the ventricle to collect at the anterior extremity of the optic lobe, and then course sharply downwards and forwards to reach the surface of the brain. The optic chiasma is a simple crossing without any intermingling of fibres, so that the nerve to the right eye, for example, arises exclusively from the left side of the brain. As in *Menidia* the nerve to the left eye is uppermost at the crossing. Each optic nerve, as is usual in Teleosts, consists of a thin wide ribbon so thrown into longitudinal folds as to form a round nerve, and each exhibits $3\frac{1}{2}$ folds.* If the optic nerve could be flattened out the width of the ribbon would

---

* The number of the folds increases with the size of the nerve, judging from our sections of young plaice at different stages, and also from the condition in the adult (see fig. 28).

be about $\frac{3}{4}$ of the maximum thickness of the body. **For a** time the right optic nerve lies directly under the left, and both immediately under the right bulbus olfactorius. The right passes straight out to its eye, but the left curves over towards the left side. Both reach the eye at about the same level, perforate the sclerotic and retina, and spread over the concave surface of the latter in the usual way.

Owing to the fact that the left eye is situated over the right, the optic chiasma is less emphasised in the Plaice than in a symmetrical fish.

We now proceed to the description of the eye muscle nerves (fig. 23), taking them in their numerical order.

Nervus Oculomotorius—III.

The nucleus of the third nerve (iii.) lies dorsally on the floor of the mesocoele very near the middle line, and mostly just over the fasciculus longitudinalis dorsalis. There is no crossing as in the case of the patheticus. The fibres of the right oculomotor curve round the fasciculus and pass backwards and downwards through the brain substance, to emerge as a large nerve on the ventral surface of the brain just above the lobi inferiores. Immediately it leaves the brain the nerve takes a sharp turn forwards, and in due course fuses with the patheticus. It passes downwards on the outer side of the lobus inferioris and between this and the v.-vii. complex. After liberating the patheticus again it courses downwards inside the skull, passes through the meninges, and enters the cup-shaped cavity formed by the parasphenoid and known as the eye muscle canal. Here it divides into a smaller upper and a larger lower nerve. Now the oculomotor consists mostly of large and well myelinated fibres, but it also contains some small lightly staining fibres. These are for the most part handed over to the lower nerve, and collect at its outer side. Finally they pass over into the ciliary ganglion and hence form the radix brevis (fig. 26, *rx. b.*) of that ganglion.

The upper division of the oculomotor (*r.s.*) passes into the Rectus superior muscle of the eye. Its small fibres are distributed to the smaller fibres of its muscle.

The lower division soon after leaving the ciliary ganglion, to which it has been closely opposed, passes sharply downwards and forwards accompanied at first by the ramus ciliaris brevis from the ciliary ganglion (fig. 26, *cil. b.*). It splits into three almost equal branches, which soon take up the following positions in the vertical plane, and are as below :—

(*a*) Dorsal branch (*r. it.*). To rectus internus. Passes upwards and inwards and reaches the ventral surface of the lower or right optic nerve. It subsequently breaks up in its muscle between and below the two optic nerves.

(*b*) Intermediate branch (*r. if.*). To rectus inferior. Divides into three principal twigs which enter their muscle in the order shown in the figure.

(*c*) Ventral branch (*o.i.*). To obliquus inferior. Descends and crosses the palatinus vii. internally and for some distance lies just below and internal to it. It then rises, crosses the palatine again, and now lies to the inner side of the rectus inferior. From this point it courses almost straight forwards at the right side of the parasphenoid and ethmoid cartilage, and finally splits up to enter its muscle in the way shown in the figure.

As regards now the left side, it may be noted at once that the distribution of the eye muscle nerves, except those coursing far forwards like the patheticus and the branch of the third to the inferior oblique, is not much affected by the torsion of the head, since the parasphenoid and its eye muscle canal are simply rotated *en bloc* to the right along their longitudinal axes. The distribution of the nerves is therefore the same, except that those of the left side have been swung upwards so as to lie nearer the dorsal edge of the body, and hence above those of the right side. In the case of the branch to the inferior oblique this rotation has caused the left one to be situated at first much above the right. In front, however, it begins to turn downwards towards the parasphenoid, and the right one at the same time rising, they eventually take up corresponding positions at the sides of the parasphenoid and ethmoid cartilage. Finally the left turns upwards to reach its muscle in which it breaks up in much the same way as the right. Neither of the long eye muscle nerves (patheticus and the branch just described) of the left side reaches, at its final distribution, a much higher transverse level than that of the right.

If a comparison be made with the eye muscle nerves of *Menidia,* as described by Herrick, it will be seen that in the two forms the relations of the nerves are essentially the same.

Nervus patheticus s. trochlearis—IV.

The fourth nerve of the right side (iv. *o.s.*) consists of many large and a few small fibres all heavily myelinated. It has no connection with the communis vii. as described by Herrick in *Menidia.* The nucleus of the pathetic is situated dorsally close behind that of the oculomotor. The two pathetic nerves cross over the mesocoele as in all hitherto investigated vertebrates, so that, for example, the right nerve arises from the left side of the brain. The two nerves pass first backwards, then rise sharply over the mesocoele, cross, and leave the brain almost in the same section so as to lie wedged in between the axis of the brain and the optic lobe. It at once turns sharply downwards and forwards, and becomes closely opposed to the dorsum of the oculomotorius. For some sections the two nerves can be distinguished, but in front they appear to fuse completely, and cannot be distinguished even under the high power. The pathetic is, however, given off again from the dorsum of the oculomotor, passes forwards and downwards, pierces the membranous wall of the brain case obliquely in front, and breaks up in the superior oblique muscle of the eye as shown in the figure.

On the left side the relations of the nerve to the brain and for some distance in front are essentially the same as on the right side. As, however, it approaches the eyes (section 392 of chart), it begins to pass towards the lower or right optic nerve Subsequently it takes up a position above and to the left of the upper or left optic nerve, having now crossed over the top of the parasphenoid and lying distinctly to the right of the morphological middle line. The two optic nerves having dipped down the left pathetic crosses over the left optic to its right side. The left optic now turns upwards towards its eye, so that the left pathetic lies considerably below it. The latter afterwards passes upwards to the left side of the frontal bridge, and is seen below the right pathetic. It finally breaks up in the left superior oblique in much the same way as the right.

Nervus abducens—VI.

The sixth nerve (vi. *r.e.*), which consists mostly of large well myelinated fibres together with some small ones, arises from the medulla by two small rootlets some distance from the middle line. Both these rootlets have apparently a common nucleus situated far from the middle line and not far from the ventral surface of the brain. Soon after leaving the brain the abducens passes sharply downwards to reach the floor of the brain case. In front it passes downwards and forwards, perforates the meninges, enters the eye muscle canal, and at once reaches the rectus externus muscle which it supplies. The abducens is the most posterior of the eye muscle nerves (cp. chart), and on this account the two nerves exhibit practically no traces of asymmetry.

Before we can proceed to describe the trigeminal and facial nerves separately, it is necessary to interpolate an account of the roots and ganglia of the trigemino-facial complex as a whole (fig. 23).

As in Teleosts generally the fifth and seventh nerves at their exit from the brain, and also their ganglia, are so disposed that it is quite impossible to completely analyse them by dissection. Examination, however, of a series of Weigert sections enables us to do this without much difficulty. Macroscopically there are two roots to the facial nerve and one to the trigeminal, and three of the four ganglia of these two nerves are compacted together into one mass. Analysis by serial sections reveals the following facts :—

The most anterior root of the complex (*r.v.*) is that of the trigeminus. It lies, however, largely internal to and below the second root, so that it is at first not obvious on dissection, and emerges from the brain just below the cerebellum. It is the only root of the trigeminus, and consists of a general cutaneous and a motor component. The nucleus of the latter lies in the floor of the fourth ventricle, and the fibres pass right through the Gasserian ganglion first into the Truncus infraorbitalis (*t.inf.*) and then into the R. Mandibularis V (*man. v.*).   On account of the fact that the large ventral nerve emerging from the Gasserian ganglion contains lateral line and communis components from the facial, the term *T. maxillo-mandibularis*, which refers to the unsplit RR. maxillaris and mandibularis V, cannot strictly be applied to it. The cutaneous component of the trigeminus arises from the spinal v. tract, and its ganglion is the Gasserian ganglion. It is distributed to the skin of the face and operculum.

The second root of the complex (*r.1.vii.*) belongs wholly to the facial, and consists of 3 roots so closely packed together that it is difficult to separate them by dissection. These roots are the dorsal and ventral lateral line roots and the communis root. The whole arise together at the same level high up on the medulla and much higher than and external to the exit of the trigeminus. The ganglia of the lateral line roots are respectively the dorsal and ventral lateral line ganglia, and that of the communis root is the geniculate ganglion. The dorsal lateral line root splits into the Ramus ophthalmicus superficialis vii. and the R. buccalis vii., whilst the ventral lateral line root is continued into the Truncus hyomandibularis as the R. mandibularis externus vii. The communis root splits into the communis v., R. palatinus vii., the R. Posttrematicus vii. and the R. mandibularis internus vii. Although the three components in this root are very compacted they retain their individuality under the microscope. The communis root enters the brain first, and then the other two fuse and enter together behind and above it. The communis root in the brain passes at once into the fasciculus communis tract, and the fused two lateral line roots terminate in the tuberculum acusticum.

The third root of the complex (*r.2.vii.*) is also entirely facial and constitutes its motor root. It arises

behind the second root and much ventral to it. Its
nucleus lies in the floor of the fourth ventricle very near
the middle line, and the root below joins the ventral
lateral line root proximal to its ganglion as in *Menidia*.
After leaving the brain the motor root, which consists of
deeply staining heavily myelinated fibres, becomes so
confused with the anterior part of the auditory root that
their separation is difficult even with the microscope. The
auditory nerve, however, passes dorsally into the tuberculum acusticum, whilst the motor vii. enters the brain
below and immediately in front of it. The motor vii.
passes into the Truncus hyomandibularis.

Of the four ganglia of the complex (*g. v.-vii.*) only
one remains distinct macroscopically. This is the
ganglion of the dorsal lateral line root, which in front is
situated just dorsal to the Gasserian ganglion, and behind
overlaps externally the root of the trigeminus. It is
entirely intracranial and is partly shown in the chart as
the cells at the base of the R. buccalis vii. The other
three ganglia are crowded between the brain and the skull
wall and apparently form one mass also entirely intracranial except for the narrowed anterior extremity of the
Gasserian ganglion which extends outside the skull along
the R. ophthalmicus superficialis v. as far forward as section 472 (cp. chart). When these three ganglia are
examined in serial sections it is seen that the most
anterior is the Gasserian ganglion. This overlaps externally the geniculate ganglion situated behind it, which in
its turn overlaps externally the ventral lateral line
ganglion—the most posterior of the three. Although the
three ganglia form a single very compact mass, it is not
difficult to define their boundaries, even where, as in the
case of the first two, the character of their cells is much
the same. In the ventral lateral line ganglion, the cells,

as is usual in these ganglia, are very small and not crowded, being scattered among the nerve fibres.

The dorsal lateral line root, as already described, has a discrete ganglion, and a reference to the chart will show that the communis, ventral lateral line and motor roots enter the ganglionic mass above by two nerve bundles. The anterior one is the communis root and the posterior represents the ventral lateral line and motor roots fused together, the motor fibres of course having no connection with the ganglion cells. The entry of the trigeminus root into the Gasserian ganglion is described above.

Leaving the compound ganglionic mass are three large nerve trunks: (1) the Truncus supraorbitalis; (2) the Truncus infraorbitalis; and (3) the Truncus hyomandibularis—all compound trunks into which both trigeminal and facial nerves enter. It will be seen on reference to the chart that the last is formed by three nerve bundles from the ganglionic mass uniting together. The most anterior of these arises from the Gasserian ganglion and thus forms the trigeminal cutaneous vii. component of the hyomandibular trunk. In *Menidia* the cutaneous vii. is extracranial, and is formed by two bundles from the Gasserian ganglion fusing together. In the Plaice there are one large and two very small bundles —all intracranial. The middle of the three nerve bundles above is the communis root, and the posterior the fused ventral lateral line and motor roots. The motor vii., as mentioned above, joins the ventral lateral line root proximal to the ganglionic mass. At first they remain distinct, the motor vii. lying on the outer face of the ventral lateral line ganglion. Before leaving the ganglion, however, the two roots become almost too intermingled to be distinguished.

Before proceeding to describe the divisions of the vth

and viith nerves, it may be mentioned that the Gasserian ganglion is the only one to receive a prominent R. communicans from the sympathetic. There also arises from the same ganglion a motor nerve which has traversed the ganglion and passes to the M. depressor operculi (*m. d. op.*). This is the most posterior nerve passing through the trigemino-facial foramen.

The trigemino-facial foramen (represented by a ring in the chart) transmits the trigeminal nerve + the dorsal lateral line root of the facial + a communis vii. component. The jugular foramen (the posterior ring in the chart) transmits the hyomandibular trunk, comprising the remainder and greater part of the facial + a cutaneous component from the trigeminus.

The various nerve rami may now be described under the names of the nerves to which they belong.

Nervus Trigeminus—V.

1. **Nervus ophthalmicus profundus** (fig. 26, *o. pr.*).—The root of this nerve (Radix ophthalmici profundi) arises on the right side from the root of the trigeminus near the brain, and proximal to the Gasserian ganglion. It passes downwards and forwards over the inner face of the latter ganglion between it and the brain, and enters the profundus ganglion, which, though closely opposed to the inner face of the Gasserian, is absolutely distinct from it. From the profundus ganglion an apparently single nerve arises which leaves the skull cavity with the rest of the vth and becomes intimately attached to the sympathetic. We could not be certain whether a few fibres were not given off to accompany the R. ophthalmicus superficialis v., thus constituting a Portio ophthalmici profundi. The nerve from the profundus ganglion passed with the sympathetic through the skull wall again by a special small

foramen into the eye muscle canal. The two bundles separate in front, but for a time the union is so close that they could not be satisfactorily analysed. However, most of the profundus fibres proximal to the ciliary ganglion separate out as the Ramus ciliaris longus (*cil. l.*), but a few of them accompany the sympathetic to the ciliary ganglion (*cil. g.*) as its Radix longa (*rx. l.*). The R. ciliaris longus leaves the eye muscle canal in front and accompanies the right rectus superior muscle to the eye, which it enters from above. A true *Ramus* ophthalmicus profundus is therefore absent in the Plaice.

On the left side the profundus nerve only separates from the root of the trigeminus just before the latter reaches the Gasserian ganglion. The only other differences between the two sides are in matters of detail (cp. the description of the sympathetic).

The only other Teleost which is said to possess a separate profundus nerve and ganglion is *Trigla*, in which, according to Stannius, it is in exactly the same condition as in the Plaice, except that it leaves the skull cavity by a special foramen. A vestigial profundus nerve is also described by Herrick in *Menidia*. Its occurrence in the Plaice in the form described above is therefore of exceptional interest. Stannius missed it altogether in the Plaice, and hence his statement that the v.-vii. complexes of *Trigla* and *Pleuronectes* only differ in this respect is inaccurate.

2. **Ramus ophthalmicus superficialis** (*r. oph. sup. v.*). —Arises from the narrowed anterior extremity of the Gasserian ganglion, and constitutes the trigeminal portion of the Truncus supraorbitalis. It accompanies the nerve of the same name from the facial for a considerable distance, being at different places more or less intermingled with it. In front, however, it separates from the facial

and is seen as a slender nerve passing forwards over the eye and somewhat near the skin, which it supplies with general cutaneous fibres. It is not free from lateral line fibres, as shown by the little plexus innervating sense organs 3 to 5 of the supraorbital canal.

The Truncus infraorbitalis (*t. inf.*), consisting of the T. maxillo-mandibularis + lateral line and communis components from the facial, arises ventrally from the Gasserian ganglion and passes sharply downwards and forwards. It soon splits into two large nerves as follows:

3. R. **maxillaris superior** (or **R. maxillaris**—*mx. v.*).— Closely accompanied by a lateral line component from the facial, which will be described in its proper place. It consists largely of general cutaneous fibres, and possibly also transmits some communis vii. fibres. It passes straight forwards across the orbit on to the upper jaw, and gives off a cutaneous twig in front, accompanying the lateral line component, and is finally distributed mostly to the skin of the anterior part of the face.

4. R. **maxillaris inferior** (or R. **mandibularis**— *man. v.*).—May also contain a communis component. Divides at once into a smaller upper and a larger lower branch. Both give off some purely motor branches, and the nerve is thereafter continued obliquely downwards and forwards across the orbit in two sections—a smaller upper and a larger lower. The former contains a few motor fibres, the latter more of the same, the remaining fibres being of small calibre and staining very faintly. The upper one bifurcates, each half containing both the sensory and motor fibres, and terminates in the posterior region of the orbit. The lower one passes forwards, following the ventral curve of the orbit, and giving off several branches on the way, on to the lower jaw, on which it ends. At about the anterior region of the orbit it gives

off a branch below which follows the R. mandibularis externus vii. for a bit, but ultimately separates out again. Some of the fibres of the R. mandibularis are distributed to the mucous surface, and hence probably represent a communis vii. component.

Rami 2, 3 and 4 are known as the first, second and third divisions of the trigeminus, and hence its name. The profundus nerve, though associated with the trigeminus, may be a separate nerve altogether, and its relations with the fifth purely secondary.

Nervus Facialis—VII.

The dorsal lateral line root splits to form the first five of the following branches :—

1. **R. lateralis recurrens facialis** (*l. rec. vii.*).—This is the first branch to arise from the root, and is given off intracranially from the top of the ganglion. It soon gives off two twigs behind as shown in the chart, each of which enters the skull wall by a separate aperture. They unite in the skull wall, however, and after leaving it, pass backwards to reinforce the posterior division of the R. oticus. The main trunk passes upwards and forwards between the optic lobe and the skull wall, perforates the frontal at the place marked with a ring in the chart, and is distributed to pit organs along its course. The R. lateralis accessorius (= R. lateralis trigemini) is, as pointed out by Stannius, absent in the Plaice, but the present nerve undoubtedly corresponds to the lateral line fibres which accompany it in the Cod, as described by Herrick. There are a very few fine fibres in it, which may conceivably be a vestigial R. lateralis accessorius, but the bulk of its fibres certainly belong to the lateral line series.

A little distal to the origin of the above, and also intracranially from the top of the ganglion, at the place indicated by a spot in the chart, the second branch arises. This is the R. buccalis. It at once splits into a dorsal and a ventral division. The former itself divides as it passes through the skull wall above the trigemino-facial foramen into a posterior R. oticus and an anterior R. buccalis externus. The latter is the R. buccalis internus, and leaves the cranial cavity by the trigemino-facial foramen.

2. **R. oticus** (*r. ot.*).—Passes almost straight upwards and splits into two, one passing straight forwards and the other, except for the curious bend at its termination, straight backwards. The former supplies sense organ 13 of the infraorbital canal, the latter, after being reinforced as above described, sense organs 14 and 15 (the last two).

3. **R. buccalis externus** (*out. buc.*).—Passes down wards and forwards, gives off a twig to sense organ 12 of the infraorbital canal, and then courses as shown in the chart to supply sense organs 7 to 11 of the same canal. The first 7 sense organs of this canal were not enclosed at the stage at which the sections were cut, and were hence lying freely on the surface. The first three are supplied by the R. buccalis internus, but no nerves could be traced to the fourth, fifth and sixth sense organs. This is due to the fact that the outer buccal nerve, after supplying sense organ 7, becomes extremely thin, and as it coursed among the numerous pigment cells on this, the ocular, side, could not be traced. It is most probable, however, that it supplied sense organs 4, 5 and 6. Comparison with the eyeless side, where there is no pigment, does not help, as the canal is shorter and develops more quickly.

4. **R. buccalis internus** (*in. buc.*).—Leaves the skull cavity by the trigemino-facial foramen and accompanies the superior maxillary nerve. In front, opposite the infraorbital canal it divides into an upper inner buccal (*up. in. buc.*) and a lower inner buccal (*low. in. buc.*). The

K

former accompanies the superior maxillary nerve, receives a bundle of fibres from it as in *Menidia*, and ultimately supplies pit organs in the region of the nose, as in *Menidia* and *Gadus*, and also the supposed vestige of the left supraorbital canal with its single sense organ (*sup. c.''*). The latter after a very long course, during which it gave off no branches at all, was ultimately traced to sense organs 1, 2, 3 of the infraorbital canal.

5. After giving off the R. buccalis the main trunk of the dorsal lateral line root is continued forwards as the **R. ophthalmicus superficialis facialis**, forming the remainder of the T. supraorbitalis (*r. oph. sup. vii*), and which is closely associated with the nerve of the same name from the trigeminus. It is connected at its origin with the Gasserian ganglion, situated below it. Its course and relations will be seen on reference to the chart, and it is mostly concerned with supplying the 5 sense organs of the supraorbital canal.

6. **Ramus palatinus facialis** (*pal.*).—Arises intra-cranially from the geniculate ganglion proximal to the formation of the Truncus hyomandibularis. It remains within the skull until it reaches the region of the orbit. At first it passes forwards and downwards, very closely attached to the cranial sympathetic from section 536 to 494, and ganglion cells really belonging to the latter have been described as belonging to the palatinus. Subsequently the palatinus passes downwards, and enters the eye muscle canal. It leaves this canal in front and passes far in front of the brain straight across the orbit and above the superior maxillary v. nerve. In the anterior region of the orbit, where it lies over the roof of the pharynx, it turns sharply downwards. During its course across the orbit it gives off branches to the terminal buds in the roof of the pharynx.

The **Truncus hyomandibularis** (*t. hm.*) is formed by the union of three nerve bundles as described above. It contains the following four components as in *Menidia*, but the second is absent in *Gadus* : —

1. Cutaneous - - V.
2. Communis ⎫
3. Lateral line ventral root ⎬ VII.
4. Motor ⎭

Just as the Truncus leaves the jugular foramen it gives off a communis nerve which we have identified as the

7. **Post-trematicus vii.** (*post. vii.*).—This nerve after a short course through 10 sections fuses with the very large communis nerve from the glossopharyngeus known as Jacobson's anastomosis (*Jac. anast.*).* The post-trematicus vii. arises considerably ventral to and quite separately from the palatinus vii. Judging from its blood vessels and innervation we regard the pseudobranch of the Plaice as a single *hyoidean* demibranch, but whether anterior or posterior we have not been able to determine. In this we differ from Herrick,† who regards the pseudobranch of *Menidia* as a mandibular demibranch, and hence our post-trematicus vii = his pre-trematicus vii. We have, however, no space for a discussion of this question, and further it may be, as Herrick suggests, that the pseudobranch is not homologous throughout the Teleostean series. One of us has formerly maintained that Jacobson's anastomosis is really the palatinus (pharyngeus) ix.

* The post-trematicus arises from the hyomandibular trunk directly the latter issues from the jugular foramen. Stannius found it in the Plaice, and regards it as a sympathetic nerve, but this of course is an error, as the sympathetic is otherwise accounted for.

† Herrick also states that Jacobson's anastomosis of *Gadus* passes from vii. to ix. It is really of course the other way about, as we state above. See a more recent paper by Herrick (Jour. Comp. Neurol. xi., p. 194).

This it undoubtedly is in many Teleosts, but in the Plaice
it seems to correspond to palatinus and pre-trematicus ix.
fused together. Otherwise there is no pre-trematicus ix.
in the Plaice at all, or the whole of Jacobson's anastomosis
is that nerve, which is somewhat unlikely.

The result, therefore, of the fusion of these two nerves
is a large bundle formed in greater part of communis ix.,
but also to a lesser extent of communis vii. And as these
two components consist of exactly similar fibres, the final
course of each component can only be traced by degenera-
tion experiments. The combined nerve (*com. vii. + ix.*)
passes almost straight downwards on the inner side of the
large pseudobranch and divides into an anterior branch to
the mucosa of the roof and upper lateral wall of the
pharynx, and a posterior branch to the mucosa of the
ventro-lateral wall of the same. Although several of these
branches passed close to the pseudobranch, none could be
traced into it, as Herrick also finds in *Menidia*. But the
nerve fibres are of very fine calibre and difficult to follow,
and as the pseudobranch has no other nerve supply it
must derive its innervation from this source, and indeed
dissection shows that it does do so. But whether its fibres
come from communis ix. or vii., or both, must be subsequently determined. The large and well-developed
pseudobranch, however, may well explain the size of
Jacobson's anastomosis.

Near the origin of the post-trematicus some motor
branches are given off from the Truncus hyomandibularis
as in *Menidia* and *Gadus*. The truncus then passes outwards and downwards, and enters a canal in the hyomandibular bone, as in *Gadus*. Soon afterwards it gives off
behind a lateral line nerve known as the :—

8. **Ramus opercularis superficialis vii.** (*op. s. vii.*).—
This at once gives off two twigs which supply the last two

sense organs (10 and 11) of the hyomandibular canal, and is then continued at first backwards and then sharply downwards towards the edge of the operculum, to supply the opercular line of pit organs.

Below and before it bends forwards the Truncus hyomandibularis splits into two large nerves—(1) an upper one turning forwards, the R. mandibularis vii. (*man. vii.*), consisting of two components accompanying each other, a coarse fibred R. mandibularis externus vii. (*man. ext. vii.*) and a fine fibred R. mandibularis internus vii. (*man. int. vii.*), and (2) a R. hyoideus vii. (*r. hy.*) passing straight downwards.

9. **R. hyoideus** (*r. hy.*).—Consists of two components, a coarse motor and a fine-fibred general cutaneous—just as in *Menidia*, but differing from *Gadus*. As in *Menidia* the hyoideus below divides into anterior and posterior branches. Sense organ 9 of the hyomandibular canal is supplied from the hyoideus, but this bundle of lateral line fibres has previously been handed over to it from the external mandibular.

As the R. mandibularis passes forwards it gives off branches to pit organs, especially one long branch above corresponding to Herrick's nerve *m. vii. 8.* It may be mentioned that the two components forming the R. mandibularis are each easily followed by the microscope. Below a large lateral line branch is given off which supplies sense organs 6, 7 and 8 of the hyomandibular canal. In front, the two components separate out, and thereafter pursue independent courses.

10. **R. mandibularis internus** (*man. int. vii.*).—A communis nerve, situated much below the externus. Passes sharply inwards to the visceral surface, and does not rejoin the externus again as in *Menidia*, and as is usual in Teleosts, with the exception of *Cottus* according to Stannius·

11. **R. mandibularis externus** (*man. ext. vii.*).—A coarse-fibred lateral line nerve, but contains some fine fibres. In front is joined by a fine-fibred nerve from the R. mandibularis v., but this soon separates out again. Just opposite sense organ 3, the external mandibular perforates the dentary and thereafter lies over the roof of the hyomandibular canal. The anterior free portion of the external mandibular supplies the first 5 sense organs of the hyomandibular canal.

We may now refer very briefly to the statements of Stannius on the trigeminal and facial nerves of the Plaice. His analysis of the roots of these two nerves is given in great detail (pp. 23-25), and is remarkably accurate considering the methods at his disposal. His five roots correspond with ours as follows :—

First Root = our first or trigeminal root (cutaneous + motor)

Second ,, = the dorsal lateral line root ⎫ of our
Third , = the ventral ,, ,, ⎬ second
Fourth ,, = the communis root ⎭ root.

Fifth ,, − our third or the motor vii. root.

His statement that branches of the superior maxillary anastomose with the palatinus vii., and are distributed to the mucous membrane of the mouth, points to a communis component in the former nerve. He describes our nerve called the R. lateralis recurrens vii., but his statement that the R. palatinus vii. has a discrete opening in the skull is of course an error.

Nervus Acusticus—VIII. (Fig. 24.)

The ear is described with the other sense organs.

All the fibres of the auditory nerve (viii.) arise from the same region of the brain (tuberculum acusticum) as the lateral line fibres. The two sets of fibres form a very large complex in the brain easily recognisable by the large size of the fibres, and the density with which they stain. There can hardly be said to be a single root to the acusticus, its fibres becoming associated into at least two rami just before or on leaving the medulla. As in *Menidia* the very minute ganglion cells are not found on the acusticus until it breaks up into its ramuli. In front, as above described, the auditory nerve leaves the medulla in conjunction with the motor vii., and is confused with it. Its division into an anterior Ramus vestibularis and a posterior R. cochlearis is not so obvious as in other fishes, on account of the manner in which it emerges from the medulla. Its further divisions or ramuli are therefore now described.

1. **R. acusticus ampullæ anterioris** (*r. a. a.*).—Most anterior branch, and courses forwards wedged in between the motor vii. and the utriculus. It then passes upwards and outwards to the outer wall of the ampulla, the sense organ of which it enters from the front.

2. **R. acusticus recessus utriculi** (*r. r. u.*).—Somewhat diagrammatic in the figure, as it really passes backwards to the floor of the utriculus to reach its sense organ.

3. **R. acusticus ampullæ externæ** (*r. a. e.*).—Passes almost straight outwards underneath the floor of the utriculus towards the outer wall of the external ampulla direct to its sense organ.

4. **R. acusticus sacculi** (*r. sac.*).—Courses at first straight downwards at right-angles to the preceding ramulus, and then backwards and downwards internal to the sacculus to reach its sense organ. As it passes backwards it was connected with a small nerve bundle, the nature of which was not determined.

The nerve extending backwards to supply the two posterior sense organs of the ear is separated off from the

auditory nerve very early, and indeed almost arises separately from the brain. It gives off a nerve above, which
is the

5. **R. acusticus ampullæ posterioris** (*r. a. p.*).—Is at
first opposed to the external surface of the glossopharyngeus, and lies between it and the utriculus, but
exchanges no fibres with it. It turns upwards, crosses the
glossopharyngeus just as the latter is bending downwards,
and becomes attached to the outer face of the lateralis,
but again does not mingle with it. Still coursing
upwards it crosses the lateralis, curves outwards over the
top of the posterior ampulla behind, and, now lying
externally to the ampulla, bends forwards to reach its
sense organ.

The remainder of the posterior nerve is the

6. R. **acusticus lagenæ** (*r. l.*).—Passes backwards
over· the roof of the sacculus, gives off a bundle to that
part of its sense organ situated there, and then crosses
inwards and downwards to supply that part of the sense
organ situated on the inner wall of the sacculus near the roof.

The **Ramulus acusticus neglectus,** with its sense
organ, is absent in the Plaice. In other fishes the Ramus
vestibularis, or anterior root = ramuli 1, 2 and 3, whilst
the Ramus cochlearis, or posterior root = 4, 5 and 6 + the
R. acust. neglectus.

Nervus Glossopharyngeus—IX.

Contrary to the condition found in *Gadus* and *Menidia*
the glossopharyngeus leaves the medulla by only one root.
This, however, consists of two large bundles, which, on
being traced into the brain are seen to belong to the
motor and communis systems. There are no cutaneous
fibres in the glossopharyngeus. In the two fishes above,
the motor fibres leave the brain by a separate root.

The single root of the ixth (*r. ix.*) leaves the medulla much below and somewhat behind the root of the lateralis. It is situated quite by itself, and distinct from any other root. It passes almost straight backwards above and slightly to the inner side of the posterior division of the acusticus, and becomes related to the R. acust. ampullæ posterioris as above described. It then courses almost straight outwards and downwards, first between the sacculus and utriculus, and afterwards between the sacculus and the skull. It now bends forwards and downwards, passes through its foramen (represented by a ring in the chart), and enters the large ganglion (*g. ix.*) lying just outside the skull.

Before entering the ganglion, and just after leaving the foramen, the root gives off above a motor branch. This passes forwards over the top of the ganglion, and enters the R. post-trematicus, thus accounting for most of the motor fibres of the glossopharyngeus.

The peculiar course of the root first backwards and then forwards is due to the position of the ear. That is to say it passes straight backwards until it can escape outwards through the fissure between the sacculus and the utriculus behind.

The nerve arising from the ganglion is very flattened and ribbon-like, and soon splits into two large nerves—an upper R. post-trematicus and a lower Jacobson's anastomosis (*Jac. anast.*). The latter passes forwards, gives off a motor branch below, the fibres of which have previously traversed the ganglion to reach it, and finally anastomoses with the post-trematicus vii., which see for its subsequent course. The relations of the sympathetic to the glossopharyngeus are described with the former system.

R. **post-trematicus** (*post. ix.*).—Courses forwards

above Jacobson's anastomosis, then bends sharply down, crossing the latter externally, and passes backwards and downwards until it reaches the first branchial arch, where it divides into two almost equal branches. One now passes straight downwards on to the anterior or concave aspect of the arch. This is the uppermost and smaller of the two, and may be called, like the similar divisions of the RR. post-trematici of the vagus, the R. post-trematicus dorsalis (text-fig. 4). It courses forwards in this position giving off branches until it became too inconspicuous to be followed, which happened before the arch joined the first and second basibranchials. The other division (R. post-trematicus ventralis), the lower and larger of the two, after continuing backwards for a bit, bent downwards and forwards to reach the posterior or convex aspect of the arch, curving externally round the elbow formed by the junction of the epi- and cerato-branchials. It then follows the arch forwards in the same position, gives off a branch above, and ultimately reaches the junction of the first branchial arch with the basi-branchials. Thereafter it arrives at the lateral edge of the branchial isthmus, crossing forwards under the hypobranchial. In front of the latter, it turns sharply upwards, and reaches the dorsal surface of the isthmus near the lateral edge, and lying just under the mucous membrane at the side of and above the first basi-branchial. Just over the cerato-hyal it anastomoses with the first branchial division of the vagus. It is now on the tongue, and tapers down and is lost under the mucous membrane of its dorsal surface, thus reaching much further forwards than the dorsal division.

We now proceed to describe the vagus complex, and we find that this is formed by the Ramus lateralis vagi, belonging to the lateral line system, and having only a secondary and unimportant connection with the vagus, and the N. vagus *sensu stricto*. There are no intracranial branches from the vagus complex, as stated by Stannius.

R. **lateralis vagi** (*r. lat. x.*).—Arises from the tuberculum acusticum, like the auditory nerve and other lateral line nerves, but appreciably below the exit of the latter and considerably above and somewhat in front of the root of the glossopharyngeus. It also arises considerably in front of the roots of the true vagus. The root (*r. lat. x.*[1]) passes downwards and backwards, lying immediately external to that of the vagus proper, and for a time between it and the posterior division of the acusticus, as above described. It has, however, no connection with either, and passes out of the same foramen (indicated by a ring in the chart) as the rest of the vagus, but external to the latter. As in *Menidia* the lateralis consists mostly of the large strongly myelinated nerve fibres, but also has many smaller ones.

Immediately on leaving the skull the lateralis gives off the R. supratemporalis vagi (*r. st. x.*), at the base of which is a small ganglion distinct from the main lateralis ganglion. The supratemporal branch, as shown in the chart, is distributed on the ocular side to the 4 sense organs so far developed in the supratemporal portion of the lateral canal, and also to the first two sense organs in the main portion of the same, *i.e.*, sense organs 1 to 6. After giving off this branch the lateralis expands into the lateralis ganglion (*l. g. x.*). The nerve arising from the ganglion passes upwards, and divides, as in the Cod, into the R. lateralis superficialis vagi (*r. lat. sup. x.*), coursing just under the skin in the neighbourhood of the main portion of the lateral canal the sense organs of the anterior half of which it supplies, and the R. lateralis profundus vagi* (*r. lat. prof. x.*), coursing below the above, and between the dorsal and lateral musculature, far from the surface. The latter gives off no branches until it passes upwards and outwards to supply the sense organs of the posterior half of the main portion of the lateral canal. It is at intervals very closely connected with branches from spinal nerves, but no fibres are exchanged as far as we could see. The superficialis division, besides supplying its portion of the lateral canal, gives off in front and above, in fact as its first branch, the R. lateralis recurrens vagi (*l. rec. x.*). This is also a true lateral line nerve distributed to pit organs, and must therefore not be confounded with the R. lateralis accessorius (trigemini), although in some fishes its homologue appears to accompany the latter. Its course, which is very extensive and just under the skin, is shown in the chart, and it corresponds precisely to the similar branch arising from the facial, and called the R. lateralis recurrens facialis (*l. rec. vii.*).

Nervus Vagus—X.

Arises by a single large root, which is, however, formed by several large bundles (Stannius says 5) uniting just on the surface of the medulla, and which is further reinforced by two very small bundles in front, as shown in the chart. This root (*r. x.*) is quite distinct both from the roots of the glossopharyngeus and lateralis,† and in fact the roots of these three nerves are more distinct and clear in the Plaice than in most Teleosts. The vagus root

---

* These nerves must not be confused with the R. lateralis trigemini of the Cod (cp. Parker's "Zootomy"), which is not represented in the Plaice at all. The latter is a communis nerve supplying its own, and *not* lateral, sense organs. It has no connection typically with the trigeminal nerve, and should be called the R. lateralis accessorius.

† Stannius states that fibres pass from this root to the root of the lateralis in the Plaice, but this is not the case.

consists very largely of communis fibres from the Lobus vagi, but also contains motor and cutaneous components. In its intracranial course it is almost entirely covered by the root of the lateralis, and before it reaches its foramen in the skull, and whilst passing through it proximally, it bears a smallish ganglion distinctly separated from the other vagus ganglia. This is the jugular ganglion (*Jug. g.*), also found by Herrick in *Gadus* and *Menidia*. It is the ganglion of the cutaneous fibres of the vagus, and forms typically the R. cutaneus dorsalis vagi, and the R. opercularis vagi.

R. opercularis vagi (*r. op. x.*).—This is given off directly the vagus leaves the skull, and at its origin is very closely opposed to the base of the R. supratemporalis vagi (see chart), but it does not fuse peripherally with it, as in *Gadus* according to Herrick. It passes forwards, and divides into antero-dorsal and postero-ventral branches supplying the skin of the opercular and supra-opercular regions. It contains both light and heavily myelinated fibres. After giving off this ramus, the vagus swells into the large ganglionic complex (*g. x.* 2-5), of which only the first ganglion (*g. x.* 1) is completely distinct.

1.—Truncus branchialis primus Vagi (*t. x.* 1).

Arises from the dorsal aspect of the vagus on its inner surface. It passes inwards and backwards towards the first spinal sympathetic ganglion, to which it becomes closely opposed. It then bends forwards and downwards, and at once swells into its large ganglion (*g. x.* 1), which is quite distinct from the other vagus ganglia, as is general among Teleosts. The motor component of the truncus passes over the external surface of the ganglion. Distal to the latter the truncus passes downwards and forwards and divides into an upper pre-trematic + pharyngeal branch, and a lower R. post-trematicus. The former passes forwards, and divides into two rami, both of which at once turn downwards and back-wards. These are the R. pharyngeus (*ph. x.* 1) and the R. pre-trematicus (*pre.* 1). The pharyngeus courses down wards and immediately breaks up in the mucous mem brane of the roof of the mouth. The pre-trematicus reaches the first branchial arch, lying just internal to the undivided post-trematicus ix. When the latter divides it thereafter accompanies the ventral division of it, but curves round the *inner* aspect of the elbow formed by the epi- and cerato-branchials.

The R. **post-trematicus** (*post.* 1.) gives off below just at its origin a motor branch, and shortly afterwards curves downwards, outwards and backwards to reach the second branchial arch. There it divides into two branches, just as in the ixth, and these bend externally round the epi- and cerato-branchial elbow, one lying above the cerato-branchial and the other below it. Thereafter their course resembles that of the post-trematicus ix.

2.—T. branchialis secundus Vagi (*t. x.* 2).

The ganglion of this division is more or less massed with the remaining vagus ganglia (*g. x.* 2-5), and their boundaries are difficult to determine. The truncus arises from the internal surface of the ganglionic mass, like the first. A small mixed plexus of communis and motor fibres, not shown in the chart, may be described here. It arises by 3 roots—one from the second branchial ganglion, and two from the third. These three roots form a plexus from which 3 nerves arise, two of which pass forwards and inwards and break up in the roof of the pharynx, whilst the third also passes forwards as a purely motor nerve.

The truncus now courses downwards and forwards, and almost at once divides into an upper palatine + pre-trematic branch and a lower R. post-trematicus (*post.* 2). The former gives off 3 RR. pharyngei (*ph. x.* 2) to the roof of the mouth, and is then continued downwards on to the second branchial arch as the R. pre-trematicus (*pre.* 2). The latter, as it passes forwards, gives off from its upper border two motor twigs not shown in the chart, and finally breaks up as usual into two branches, which pass on to the third branchial arch, and the distribution of which is essentially the same as the corresponding divisions of the first branchial trunk.

3.—T. branchialis tertius Vagi (*t. x.* 3).

Arises from the ventral edge of the compound ganglionic mass. The two nerves it contributes to the pharyngeal plexus have been described above. Another nerve not shown on the chart arises posteriorly and internally from the base of the truncus. It passes sharply downwards and backwards to the roof of the pharynx. The truncus itself, directly it leaves the ganglion, divides into a palatine + pre-trematic branch coursing gradually downwards and forwards, and a R. post-trematicus, passing sharply downwards and backwards. The former divides as usual into a R. pharyngeus (*ph. x.* 3) and a R. pre-trematicus (*pre.* 3), the latter passing on to the third branchial arch. The R. post-trematicus (*post.* 3) breaks up into the usual two branches much sooner than usual, as in *Menidia*. The anterior one gives off at once in front a motor branch. Both pass on to the fourth branchial arch and are distributed as usual.

The T. branchialis quartus vagi is so closely associated with the remainder of the vagus that it cannot be separately described.   After giving off the branchialis tertius, the vagus passes backwards and splits into two large nerves.   The upper one is the R. intestinalis vagi (*r. intest. x.*), and the lower one contains the fourth branchial trunk, together with the RR. cardiacus et œsophageus vagi.   The upper division gives off two branches, which join the rich œsophageal plexus formed by branches of the lower division, and afterwards passes almost straight backwards as the R. intestinalis, wedged in at first between the kidney, thymus and roof of the œsophagus.   It gives off branches to the œsophagus from time to time, and passes downwards until it lies at the side of the latter structure.   It ultimately splits into two, both of which break up in the lateral wall of the œsophagus and are lost before the stomach is reached.

The lower division, before and after it separates from the R. intestinalis, gives off about 10 mostly motor nerves, which at once form an elaborate plexus in the region of the dorso-lateral wall of the œsophagus.   These nerves and others are not shown in the chart.   It then passes slightly downwards and backwards, and gives off in front the fourth R. pre-trematicus (*pre.* 4), which passes sharply downwards and forwards.   This at once gives off in front a small motor twig, and then courses straight on to the fourth branchial arch.   It was not observed to give off a R. pharyngeus unless an extremely small twig distributed apparently to the side wall of the pharynx represented that branch.

After giving off the fourth pre-trematicus, the lower division turns almost straight downwards, and then divides into four branches.   Two of these pass forwards at once into the ventral wall of the œsophagus, and represent the final derivatives of the R. œsophageus.   The third accompanies the inferior pharyngeal bone, and is therefore the fourth R. post-trematicus (*post.* 4).* The last branch continues the downward course, gives off a motor branch behind, and then courses forwards and downwards in close contact with the roof of the pericardium. This may be the R. cardiacus (*r. car.?*), but it does not actually pass on to the sinus venosus, and certainly contains a number of motor fibres distributed outside the heart. The true R. cardiacus may therefore have been overlooked, especially as Stannius proved by stimulation experiments that the vagus of the Plaice sent fibres to the heart.

3.—THE SPINAL NERVES (Fig. 27).

The Fourth Spinal Nerve.

We describe this spinal nerve first since in most respects it may be taken to represent the structure of most of the other spinal nerves. The visceral fibres are not taken into account.

The fourth spinal nerve arises by two roots—a dorsal largely sensory (*d.* 4) and a ventral motor (*v.* 4). Each root leaves the neural arch of the third vertebra by a separate foramen, as described by Stannius, and passes at once into a single large extra-vertebral ganglion (*g.* 4). The motor fibres for the R. medius and R. ventralis perforate the ganglion, but those for the R. spinosus pass upwards internal to it. One lateral, one ventral and two dorsal branches arise from the nerve. The two last are the R. communicans (sensory) and the R. spinosus (motor), whilst the former are respectively the R. medius (sensory and motor) and the R. ventralis (sensory and motor).

1. **R. communicans** (*r. com.* 4).—This sensory ramus

* We differ from Herrick in naming two of the branches of the Vagus. His fourth post-trematic (fig. 4, post 4) is apparently the pharyngeus iv., and his " branches of the Vagus for the inferior pharyngeal teeth " (ph. v., same figure) are equivalent to our post-trematic iv,

arises from the antero-dorsal region of the ganglion, passes upwards, and fuses with the R. spinosus of the nerve in front (third spinal).

2. R. spinosus (*r. sp.* 4).—A motor ramus leaving the postero-dorsal region of the ganglion, and giving off near its base a posterior lateral branch for the intermediate portion of the dorsal musculature. It then passes dorsally, and fuses with the R. communicans of the nerve behind (fifth spinal). The resulting mixed trunk continues upwards and forwards in the dorsal musculature very close to the middle line, and is distributed to the dorsal portion of the dorsal musculature, the inter-spinal muscles and the dorsal skin.

3. R. medius (*r. m.* 4).—This mixed ramus, leaving the ventral extremity of the ganglion, courses laterally outwards through the ventral portion of the dorsal museulature, and bifurcates. The upper division accompanies the intermuscular bone, and supplies the ventral portion of the dorsal musculature. The lower division passes downwards into the lateral musculature (which it sup plies), obliquely crossing under the R. lateralis profundus vagi (fig. 23, *r. lat. prof. x.*), to which it may be very closely attached, but with which it never mingles. The sensory fibres of the ramus pass out laterally to the skin, and supply that portion of it around the lateral sensory canal.

4. R. ventralis (*r. v.* 4).—This mixed ramus also arises from the ventral extremity of the ganglion, and is the largest of all. It passes downwards, and just over the kidney receives the R. communicans from the fourth spinal sympathetic ganglion (*com. iv.*). It then turns outwards between the lateral musculature and the kidney, and afterwards downwards again between the lateral musculature and the liver. Finally it continues down-

wards on the inner surface of the abdominal wall, and just under the peritoneum. This ramus supplies the ventral musculature and ventral skin. In the region of the appendages the limb girdles and fins are supplied from R.R. ventrales. In the specimen now investigated, however, the fourth spinal was not connected with either the pectoral or pelvic appendage. The fourth R. ventralis anastomoses below with the nerve $r.$ $v.$ $2 + 3^1$, as described below.

The First Spinal Nerve.

This nerve is a compound of at least two spinal nerves, since it has two ganglia and most of its principal rami are in duplicate. It is, however, here described as the first spinal, on account of the difficulty of completely isolating its constituents.

The ganglia and roots of the first spinal are situated in the bony tube formed by the exoccipital, and leading from the foramen magnum into the cranial cavity proper (see fig. 4). There is one main foramen for the nerve, which tunnels transversely the narrowed base of the paroccipital condyle, as shown in fig. 3 (above the lower letters *Ex. O.*). Another smaller foramen for the R. spinosus b. is situated immediately above this, and occasionally there is another larger one immediately below it for the R. ventralis, as shown in the chart. Usually, however, the latter nerve passes through the main foramen. There are thus at least two foramina for the first spinal nerve, and there may be three—all situated in the exoccipital.

The first spinal has two perfectly distinct ganglia— an intracranial ganglion (*g. intcr.*), and an extracranial ganglion (*g. extcr.*). These two ganglia are connected by a large bundle of sensory and motor fibres which pass through the main foramen (indicated in the chart by the oval dotted area). There was in the specimen sectioned a discrete patch of cells on the right side on the R. ventralis also, but in other examples these were continuous with and part of the extracranial ganglion.

We follow Fürbringer and Herrick in designating the cephalic constituent of the first spinal by the letter *b*, and the caudal constituent by the letter *c*. It will be observed that the Plaice has three ventral roots instead of the two described by Herrick in *Menidia*,* and of these the extra one is undoubtedly the first (*v. b.*¹).

The first spinal nerve of the Plaice has two dorsal sensory (mostly) and three ventral motor roots. Of the two dorsal roots the first (*d. b.*) is larger than the second (*d. c.*), and arises obviously from the spinal vth tract. They both pass into the intracranial ganglion. Of the three ventral roots, the first (*v. b.*¹) is very long and slender, and fuses with the second (*v. b.*). The third (*v. c.*) is very short, and is the largest of all the roots. All three ventral roots pass into the intracranial ganglion.

The following nerves arise from the intracranial ganglion :—

1. **R. spinosus, b** (*r. sp. b.*).—A motor nerve, arising from the fused first and second ventral roots. It passes through the intracranial ganglion, and leaves the exoccipital by a small foramen immediately above the main foramen (indicated by a ring in the chart). It then passes forwards over the top of the extracranial ganglion, rises sharply at the side of the auditory capsule, and afterwards turns forwards over the roof of the capsule to supply the dorsal musculature and interspinal muscles. This nerve does not anastomose with a sensory R. communicans like the posterior RR. spinosi.

---

* Stannius mentions only four roots in the Plaice also, having apparently missed the first ventral.

2. **R. spinosus, c** (*r. sp. c.*).—Large motor nerve from the third ventral root. It perforates the intracranial ganglion dorsally, passes a little backwards in order to leave the skull by the foramen magnum above, and then courses forwards and upwards. Arrived at the roof of the skull, it bends forwards over the latter, at first lying over the epiotic a little to the side of the narrowed posterior portion of the supraoccipital. It then fuses in the typical manner with the sensory R. communicans of the second spinal nerve, to form a conspicuous mixed nerve which courses forwards over the roof of the skull near the middle line to its distribution.

3. **R. ventralis** (*r. v. b + c*).—This usually leaves the skull by the main foramen, but it occasionally has a foramen of its own situated below the main foramen, and above the paroccipital condyle, as in the specimen figured. It has a comparatively slight connection with the extra-cranial ganglion, but the latter ganglion does undoubtedly contribute fibres to it. The R. ventralis is formed as follows:—First of all the remainder and greater part of the first and second ventral roots, having passed underneath the intracranial ganglion, and together with some sensory fibres, pass into the special foramen (indicated by a ring in the chart). They are immediately followed by the remainder of the third ventral root (also accompanied by sensory fibres from the intracranial ganglion), which, having perforated the intracranial ganglion, and instead of passing into the main foramen as usual, turned downwards and entered the special foramen. The ventral root thus left the skull by all three foramina. These two mixed trunks more or less unite in the foramen, and immediately outside it in this specimen bore a small number of discrete ganglion cells. The latter, however, undoubtedly belong to the extracranial ganglion. The two trunks soon

separate again into two mixed nerves, and pass downwards and backwards. The lower one receives the R. communicans from the first spinal sympathetic ganglion (*com.* 1). They subsequently unite again, and distal to the union a large sensory and motor nerve is given off, which curves downwards and forwards. This is the R. cervicalis or so-called N. hypoglossus (*r. cerv.*). It courses at first at the side of the pericardium opposite the junction of the auricle and ventricle, and the upper sensory and the lower motor components are quite obvious. In front, the two components separate out into a dorsal sensory and a ventral motor nerve. Shortly after giving off the above, the R. ventralis receives a sensory and motor anastomosis from the R. ventralis of the second spinal nerve (*r. v.* 2^1), thus forming the brachial plexus. The compound trunk (*r. v.* 1) then courses downwards and slightly backwards, and gives off a motor nerve in front and behind to the muscles of the pectoral girdle (*r. v.* 1^1 and *r. v.* 1^2). It now passes through the scapular fenestra (see fig. 8—indicated by a ring in the chart), in order to reach the external aspect of the pectoral girdle, where the two components at once separate out into an anterior motor nerve (*r. v.* 1^3) and a posterior sensory nerve (*r. v.* 1^4), which are distributed to the pectoral fin, the latter to its dorsal region.

The following nerves arise from the extracranial ganglion :—

1. R. **communicans, b.** (*r. com. b.*).—A large sensory nerve from the extreme dorsal point of the ganglion. It passes forwards over the roof of the skull at the same transverse level as the R. spinosus, b., but external to it. It ultimately passes first outwards and then downwards between the skull and the skin, and is distributed to the latter in the region of the auditory capsule. As the R.

lateralis accessorius is absent in the Plaice, and as it cannot anastomose with the R. spinosus, this nerve is not an actual R. communicans in the Plaice.

2. **R. medius, b.** (*r. m. b.*).—Leaves the ganglion as two distinct strands—an anterior sensory and a posterior motor. The sensory section is small and passes outwards and forwards to the skin in the region of the lateral sensory canal. The motor section arises from the fibres of the fused first and second ventral roots which have passed underneath the intracranial ganglion, traversed the main foramen lying underneath motor fibres from the third ventral root, and perforated the extracranial ganglion. Immediately it leaves the latter ganglion it gives off a small anterior branch, whilst the rest of the nerve passes laterally backwards and breaks up largely in the dorsal musculature.

3. **R. medius, c.** (*r. m. c.*).—A large mostly motor nerve arising almost entirely from the third ventral root. The fibres pass through the main foramen, and perforate and leave the extracranial ganglion at its anterior aspect. Directly it leaves the ganglion, it gives off above a small sensory thread, which passes outwards and backwards towards the lateral sensory canal. The remainder and larger part of the nerve courses downwards, outwards and backwards in the dorsal musculature, in which most of its fibres break up, except also a few that pass outwards towards the skin.

The Second Spinal Nerve

The second spinal nerve has a dorsal mostly sensory (*d.* 2) and a ventral motor (*v.* 2) root, and two extravertebral ganglia—a large dorsal ganglion (*g. d.* 2) and a smaller ventral ganglion (*g. v.* 2). The dorsal root passes backwards at once into the dorsal ganglion, but the ventral root passes upwards and splits into two bundles, one penetrating the dorsal, the other the ventral ganglion. Both roots emerge from the atlas vertebra by a single foramen (fig. 17).

The following nerves arise from the dorsal ganglion:

1. **R. communicans** (*r. com.* 2).—A sensory nerve arising from the extreme dorsal tip of the dorsal ganglion. Passes upwards and forwards, bends forwards over the roof of the skull, and fuses with the R. spinosus c. of the first spinal nerve (*q. v.*).

2. **R. spinosus** (*r. sp.* 2).—A motor nerve passing upwards internal to the dorsal ganglion. After giving off a fine motor twig below, which passes upwards external to the dorsal ganglion (and not shown in the chart), it liberates in front and above a larger motor nerve which at once splits into two bundles coursing laterally in the dorsal musculature—one anteriorly and the other posteriorly (see chart). Above, it receives the sensory R. communicans from the third spinal nerve in the typical manner, and bends forwards over the roof of the skull, keeping very close to the middle line.

The following nerves arise from the ventral ganglion:

1. **R. medius** (*r. m.* 2).—A mostly motor nerve, which perforates the ganglion, and then turns laterally backwards in the dorsal musculature. It breaks into two—one coursing in the dorsal musculature above the R. lateralis profundus vagi, and the other crossing below it into the lateral musculature, and supplying the muscles in these regions. It seems to contain some sensory fibres also.

2. **R. ventralis** (*r. v.* 2).—A mixed nerve perforating the ganglion and coursing downwards and backwards over the kidney and approximating to the R. ventralis of the first spinal nerve. It receives two Rami communicantes

from the second spinal sympathetic ganglion (*com. ii.*), and sends a mixed bundle to the R. ventralis of the first spinal nerve as above described (*r. v.* 2^1). Just at about the same place it fuses completely with the R. ventralis of the third spinal nerve, but the above anastomosis is derived from the R. ventralis 2, and contains no fibres from 3. The compound trunk (*r. v.* $2+3$), after giving off a small branch to the inner surface of the clavicle (not shown in the chart), courses forwards and downwards, and gives off below two motor branches which pass at first downwards and then forwards to supply the ventral musculature. They anastomose with each other close to their origin, and the posterior one (*r. v.* $2+3^1$) also anastomoses below with the R. ventralis of the fourth spinal nerve. The remainder of the trunk then splits behind into two—a small motor nerve and a larger mostly sensory nerve. The latter (*r. v.* $2+3''$) is continued backwards on to the ventral portion of the pectoral fin.

The Third Spinal Nerve

This spinal nerve has the usual two roots (*d.* 3 and *v* 3), and a single very large extra-vertebral ganglion (*g* 3). Each root has a separate foramen in the neural arch of the second vertebra (fig. 17). The sensory R. communicans (*r. com.* 3) contains a few motor fibres, which are liberated as a small bundle for the dorsal musculature (not shown in the chart). It fuses with the R. spinosus 2, as above described. The motor R. spinosus (*r. sp.* 3), after rising internal to the ganglion, courses at first backwards, fuses with the R. communicans 4, and then turns forwards as a mixed nerve high over the roof of the skull and very close to the middle line. It gives off below the motor nerve to the dorsal musculature just like the preceding nerve.

The **R. medius** (*r. m.* 3) is a mixed nerve, but largely motor. It breaks up into the two branches as in the second spinal, the upper one for a time accompanying the intermuscular bone, and the lower the R. lateralis profundus vagi (but not mixing with it).

The **R. ventralis** (*r. v.* 3) is also a mixed nerve. It receives the R. communicans from the third spinal sympathetic ganglion (*com.* III), and then fuses with the R. ventralis of the second spinal, as above described.

It is thus seen that the pectoral girdle and fin of this specimen of the Plaice is supplied largely by the first spinal, but also to a certain extent by the second and third. It must, however, be emphasized that the limb plexuses are subject to great variation, that the above account is not a generalised description, and therefore takes no account of variations. Stannius, however, states that fibres from the RR. anteriores of the first three spinal nerves pass to the pectoral fin in the Plaice.

With regard to the fifth spinal nerve, we need only mention that the R. medius arises from the R. ventralis, and that its lower division appears to completely fuse with the R. lateralis profundus vagi, but really is only very closely attached to it.

The innervation of the paired fins of Teleosts has very important theoretical bearings. In the Plaice from which the spinal nerves were plotted out, the R. ventralis of the fifth spinal nerve, together with that from the sixth,* were the nerves which supplied the pelvic fin. Now if this fin is homologous throughout Teleosts generally,

---

* Stannius states that the pelvic fin of the Plaice is innervated from the fourth and fifth spinal nerves, and this tallies with Cuvier's scheme. In our sections, however, the fourth spinal nerve was not connected with the pelvic fin.

which we assume will not be doubted, then we must con-
clude that we have, in this clearly monophyletic group, a
genuine case of fin migration, since the pelvic fin of
Teleosts may be either abdominal, at the side of the anus,
thoracic, just behind the pectoral fin, or jugular, in front
of the pectoral. Further the fin must have moved from
behind forwards, since it may be clearly deduced from the
fossil forms that the abdominal position is unquestionably
more primitive than the jugular. Now in Elasmobranch
fishes, according to Gegenbaur's hypothesis, the pelvic fin,
having been formed from the branchial skeleton, must
have migrated from before backwards, unless the addi-
tional hypothesis of the extension of the branchial region
further back than we have any knowledge, is evoked.† It
must be noted at once, first, that all palæontological evidence is against such an assumption, and second, that this
migration, if it occurred at all, must have taken place at
a remote geological period. On the other hand, the
migration of the pelvic fin of the Teleost is not only something more than an hypothesis, but it must also have
occurred within comparatively recent times. Now several
attempts have been made to deduce the migration of the
Elasmobranch fin from its innervation, and so far the
Teleosts have been ignored, in spite of the fact that here
we might with much more reason expect to encounter
such evidence. Unfortunately, however, in the latter
group, the jugular pelvic, as in the case of the Plaice, is
supplied by the anterior spinal nerves of its region, and
the abdominal pelvic, as in the several fishes investigated
by Stannius, is supplied by the posterior nerves of its

---

† It may, of course, be maintained, as it is in the case of some Elasmobranchs, that the pelvic fin of Teleosts migrated *first* backwards and *then*
forwards. We have ignored the former possibility (which after all would
be purely hypothetical), in order to concentrate attention on the latter,
where the migration theory may be the more satisfactorily tested.

region.   Here, therefore, the innervation clearly proves
nothing, and it may at least be doubted whether it proves
any more in the case of the Elasmobranchs.   We refer
now purely to neurological evidence, and are of course
aware that the supposed migration of the Elasmobranch
fin is alleged to be supported by ontogeny also.

Stannius (op. cit.) makes several references to the
spinal nerves of the Plaice, and he quite realised the
morphological value of the anterior extremity of the dorsal
fin, since he carefully distinguishes between the forward
extension of mixed spinal nerves over the roof of the
cranium, and the sensory dorsal branches of the cranial
nerves themselves.   His work also contains a figure of the
anterior spinal nerves of the Plaice (Taf. iv., fig. 1).

4.—The Sympathetic Nervous System (Fig. 26).

Owing to the impossibility of satisfactorily dissecting
the anterior portion of the sympathetic, we plotted it out
from the same series of sections as were used in the case
of fig. 23.   As it is drawn to the same scale, the two figs.
may therefore be compared.   Our examination of the
sympathetic commenced at the sixth vertebra, and behind
the seventh spinal nerve.   It is described from behind
forwards, right side first.

The sympathetic cord may be conveniently divided
into two parts, which we will call the cranial sympathetic,
associated with the skull and cranial nerves, and the
spinal sympathetic, associated with the vertebral column
and spinal nerves.   In each portion, the ganglia are
numbered separately from before backwards.

At the sixth vertebra the sympathetic is situated
laterally below the centrum and the transverse process.
It here sends a short Ramus communicans to the ventral
ramus of the seventh spinal nerve (com. vii.), which

crosses it at right-angles. In front of this is found the seventh spinal sympathetic ganglion (7'). At this region the two cords are connected by a transverse commissure below the dorsal aorta and bearing a pair of ganglia (7''), formed as shown in the figure, and the cord of the right side is also looped. There are two RR. communicantes to the ventral ramus of the sixth spinal nerve (*com. vi.*), but in front of this, beyond a loop for a renal vein and a very small ganglion in front of ganglion 5 (5'), there are no features of special interest until we come to the second spinal sympathetic ganglion.

The sympathetic behind the second ganglion lies immediately above the inner dorsal angle of the kidney. The coeliac ganglion (*g. coel.*) lies close under the second spinal sympathetic (2'), and in front rises up to fuse with it. Four nerves arise from the second ganglion. Behind, a pair (*com. ii.*) pass independently into the ventral ramus of the second spinal nerve, and thus form RR. communicantes ii. The sympathetic now lies internal to the kidney, and just above the cœliaco-mesenteric artery. In front, a large third nerve arises dorsally, and soon splits into an anterior and a posterior branch. The former is continued into the first ganglion (1') and thereafter into the cranial sympathetic, whilst the latter forms a prominent ganglionated (2'') commissure under the dorsal aorta with the cord of the other side. The fourth branch arises anterior to the third, and curved backwards on to the aorta, where it was lost. The second ganglion now tapers down, and terminates in close proximity to the kidney.

The coeliac ganglion, which it should be noted arises from the *right* sympathetic cord, passes backwards after fusing with the second ganglion, and is situated just over the coeliaco-mesenteric artery. It also gives off four branches. The most dorsal one passes straight into the kidney after a very short course. Of the three others two join to form the dorsal Nervus splanchnicus (*n. sp.'*) lying over the above artery, but before joining one of them gives off a thin twig which accompanies the œsophageal artery. The fourth branch, after giving off a fine twig to the kidney, becomes the ventral N. splanchnicus (*n. sp."*), lying under the artery. The latter nerve is for some part of its course opposed to the R. intestinalis of the vagus. When the coeliaco-mesenteric artery divides into a dorsal coeliac and a ventral mesenteric artery, the two RR. splanchnici also divide (cp. figure).

From the first spinal sympathetic ganglion (1') three nerves arise. One of these is certainly, and another possibly, a R. communicans to the ventral ramus of the first spinal nerve (*com.* 1). The third seemed to enter the jugular ganglion of the vagus, but this is not certain. From sections 730 to 704 the cord is attached to the inner surface of the root and ganglion of the Truncus branchialis primus N. vagi, but as far as could be ascertained exchanged no fibres with it. The first spinal sympathetic ganglion occupies an intermediate position between the cranial and spinal sections of the cord, since it is connected both with the vagus and first spinal nerve.

After leaving the first spinal ganglion, the cord is continued forwards as the cranial sympathetic, and from section 702 to 678 accompanies the superior jugular vein, during which it bears a very small ganglion (8). After leaving this vein it bears ganglion 7 and becomes attached to the inner surface of the glossopharyngeus ganglion and trunk (658-620), being wedged in between the latter and the skull, and bearing ganglion 6. In front again, when the glossopharyngeus splits up, it accompanies the ventral edge of Jacobson's anastomosis (618-560), and swells into a very large ganglion (5) very closely related to Jacobson's anastomosis. On leaving the latter it bears another moderate-sized ganglion (4), but no fibres were seen to be exchanged between any of these ganglia and the glosso pharyngeus nerve. In front of ganglion 4, the cord rises upwards and becomes attached to the Truncus hyoman dibularis, just as the latter emerges from the skull, the post-trematicus vii. lying below it. It now passes into the cranial cavity through the jugular foramen, and is so closely pressed against the T. hyomandibularis that we were unable to determine whether fibres were exchanged or not, although we believe not. Inside the cranium, it bears the small ganglion 3, from which the ganglionated intracranial most anterior commissure (3″) arises. The commissural ganglion is situated actually on the root of the sixth cranial nerve.

From section 536 to 494 the cranial sympathetic accompanies the R. palatinus facialis, and is very closely attached to it. In front, it passes through the trigemino-facial foramen, and takes up a position between the skull and the origins of the maxillary and mandibular v. nerves. It now swells into the large ganglion 2, from which a very prominent R. communicans (*com. v.*[1]) passes upwards to the T. maxillo-mandibularis. The first cranial ganglion (1) lies above the second, and is connected with it by a very short strand of fibres. Both the first and second ganglia give off a cord in front, but the two unite before reaching the ciliary ganglion. The first ganglion also gives off a nerve internally, which accompanies the ventral edge of the R. ophthalmicus superficialis v. The possibly corresponding nerve of the other side arises externally from the second ganglion, accompanies the superior maxillary v., and has a very small ganglion of its own.

From the first cranial ganglion onwards the sym-pathetic is accompanied by the profundus nerve. They pass together through a special foramen in the ventral edge of the alisphenoid into the eye muscle canal. The cord from the second ganglion also enters the canal by the same foramen. Before reaching the ciliary ganglion the profundus nerve separates from the sympathetic as the Ramus ciliaris longus (see profundus nerve), but some fibres are dispatched from it to accompany the sympathetic to the ganglion as the Radix longa. The ciliary ganglion itself (*cil. g.*) is closely attached above to the main trunk of the oculomotorius, after the latter has given off the nerve to the rectus superior. The Radix brevis therefore is exceedingly short (*rx. b.*). From the ciliary ganglion in front arises the R. ciliaris brevis (*cil. b.*). This passes forwards in the eye muscle canal, accompanying the main trunk of the oculomotorius, until the latter breaks up. It then courses under the lower or right optic nerve as a conspicuous bundle, enters the eye ball with it, and thereafter passes downwards and forwards towards the iris.

As regards now the eyeless or left side, the sympathetic nervous system, like the nervous system generally, is not so well developed. This is especially noticeable in the ganglia, which are perceptibly smaller than those of the other side. In front, the disturbance of the symmetry has dragged the sympathetic over to the ocular side. Generally speaking, however, the left side resembles the right in all essential respects, but the following differences may be mentioned.

In the spinal sympathetic the seventh ganglion (7') is in two parts, R. communicans vi. (*com. vi.*) is separated from its ganglion (6'), and R. communicans v. (*com. v.*) is situated between a large and a small ganglion (5' and 5''). Ganglion 2 gives off externally a large nerve which passes backwards and downwards to the kidney. R. communicans ii. (*com. ii.*) is single, but the first (*com. i.*) is however double, the posterior of which bears a very small ganglion. No exchange of fibres between the sympathetic and the vagus was observed.

In the cranial sympathetic only 7 ganglia were found instead of 8. When the cord reaches the glosso pharyngeus, instead of becoming attached to the ventral border of the ganglion and subsequently to that of Jacobson's anastomosis, it passes upwards internally to the ixth and becomes opposed to its upper division. This is due to the fact that when the ixth splits it forms a *dorsal* Jacobson's anastomosis and a *ventral* post-trematicus, instead of the reverse as on the right side, and the sympathetic always accompanies the former. Ganglion 5 lies quite clear of and above the glossopharyngeus, in striking contrast to the condition on the other side. As the cord passes through the jugular foramen it is for a time very tightly wedged into the angle formed by the outgoing post-trematicus vii. and the hyomandibular trunk. We could not determine whether there was any exchange of fibres between the sympathetic and the facial nerve, but if present it is not obvious. There is a large R. communicans (*com. v.'*) to the base of the T. maxillo-mandibularis.

The combined sympathetic and profundus nerve when they enter the eye muscle canal lie just above the ciliary ganglion. Instead, however, of passing straight down to the ganglion, as on the right side, they curve round the left rectus externus muscle, and describe an almost complete circle before reaching the ganglion. The few fibres forming the Radix longa and the sympathetic join with the fibres leaving the ciliary ganglion to form the R. ciliaris brevis. It is doubtful whether many of them enter the ganglion at all on this side. There are no other differences of importance between the two sides.

M

We are unable to agree with some of the remarks of Stannius on the sympathetic of the Plaice. He states that the first cranial ganglion is that connected with the facial nerve, that there is no ganglion corresponding to the glossopharyngeus, and that the rami communicantes for the first 3 or 4 spinal nerves arise from a common ganglion also giving origin to the NN. splanchnici. We have not examined the sympathetic posteriorly, but Stannius states that there are very large sympathetic nerves connected by a commissure perforating the kidney to reach the reproductive organs. They also pass backwards with the ovary or testis into the cavity between the skeleton and the skin formerly supposed to be the posterior extension of the body cavity.

The sympathetic nervous system of Fishes has recently been investigated in some detail by Jaquet* and C. K. Hoffmann†—the former studying its anatomy and the latter its development. Jaquet divides it into cephalic, abdominal and caudal portions. The first is stated to be connected with the ganglia of five cranial nerves—the "hypoglossal" [first spinal], vagus, glossopharyngeus, facialis and trigeminus, and fibres from the second and third ganglia are said to accompany the glossopharyngeal to the pseudobranch. Jaquet's work contains a formal scheme of the Teleostean sympathetic (fig. 4), and also many statements which are not borne out by our examina tion of the Plaice, and which seem to us to require confirmation. The most important work on the anatomy of the sympathetic in bony fishes is that of Chevrel,‡ who investigated the relation of the ganglia to those of the cranial nerves, and who asserts that the first sympathetic

---

* Bull. Soc. Sci. Bucarest-Roumanie, Ann. x., 1901.

† Verhand. K. Akad. Wetens. Amsterdam, Sect. ii., Dl. vii., 1900.

‡ Arch. Zool. Expér., Ser. ii., T. v. bis.

ganglion is always associated with the trigeminus, except in the Physostomi, where there are no ganglia in front of the vagus. This is what we should expect, as the cranial sympathetic appears for the first time in the bony fishes, and the Physostomi are undoubtedly a primitive group of Teleosts.

F.—THE SENSE ORGANS.

1.—The System of Lateral Sense Organs or Sensory Canals.

(Figs. 23 and 29.)

The obvious line seen at the side of the body in most Fishes, including the Plaice, and known as the lateral line (Seitencanal of German authors), is in our type a long tube or lateral canal, protected by a row of modified scales (lateral line ossicles), and opening on to the surface by pores at more or less regular intervals. These pores may open directly into the canal or they may do so by the intermediation of a little tubule. At first the only substance found in such a canal as this was a quantity of a jelly-like mucus, and hence these canals were called mucous canals, and were supposed to secrete the mucus on the surface of the body. The discovery, however, that they contained large sense organs, one of which usually occurred between two succeeding surface pores, and that the mucus was only of minor importance, and developed by the numerous goblet cells in the lateral canal itself to enable the sense organs to perform their function, and also that the mucus in the lateral canal was of quite a different character to that occurring on the surface of the body, conclusively proved that these lateral canals constituted a *sensory* structure and could not therefore be called mucous canals. We hence adopt the term of sensory canals originally proposed by Ewart.

Now the sensory canals are not confined to the body of the fish, but always extend on to the head, where, as a very general rule, they form a much more complex and important system of sensory canals, which, on account of their deeper situation and therefore less obvious character, are too often disregarded. In the Teleostean Fishes the innervation of this system is so remarkably constant in those forms in which it has been properly investigated as to justify its division on each side of the body into the following four canals, the extent of any one of these being determined by its innervation. That is to say, that part of the cephalic system of sensory canals called the infra-orbital canal is precisely that part of the system innervated by a single perfectly definite nerve—the R. buccalis facialis. One extremity of this canal is a natural blind extremity, the other is the artificial boundary beyond which its nerve does not extend. The four canals are : — (1) the lateral canal at the side of the body ("lateral line" of systematists), defined by the branching of the R. lateralis vagi; (2) the supraorbital canal over the eye, defined by the R. ophthalmicus superficialis vii.; the infraorbital canal under the eye, defined by the R. buccalis vii.; and the hyomandibular canal on the oper-culum and lower jaw, defined by the R. mandibularis externus vii. These nerves are in this work only provisionally associated with the vagus and facial nerves, as their true morphological value cannot be discussed in a general treatise of this nature.

Although much work has been done on the use of this undoubtedly sensory apparatus, its function is even yet a subject for speculation. This is due to the fact that owing to its very diffuse nature and the intimate relations of its nerves with the other cranial nerves, it **has** so far been found impossible to devise really satisfactory experiments. F. S. Lee, the latest investigator of the function of the lateral line organs, concludes that they are connected with the sense of pressure or equilibration, and even if this be not the case, there is some reason to doubt whether the latter can be located in the semi-circular canals as in higher vertebrates.

The Lateral Canal (*lat. c.*).—This canal, in part the "lateral line" of systematists, is supported during the greater part of its length by modified scales. Arrived at the region of the shoulder girdle it tunnels through the post-temporal, and then for a short distance has no bony support. It now enters a chain of ossicles, undoubtedly metamorphosed scales, called supratemporal ossicles or extrascapulæ. The last or most posterior supratemporal is attached to the dorsal surface of the pterotic at the region of the posterior depression shown in fig. 1, and is larger than any of the others. It consists of 2 parts, one running longitudinally in a curve for the "lateral line," and the other transversely and somewhat forwards for the terminal portion of the supratemporal canal. Whilst in this ossicle the lateral canal on the ocular side anastomoses with the posterior extremity of the infraorbital canal, and then turns abruptly upwards almost at right-angles, and afterwards forwards, the latter or anterior portion of the canal being called by some authors the **supratemporal canal** (*s. t. c.*). The last supratemporal ossicle of the Plaice therefore corresponds to the second, third and fourth of the Cod fused together. In one Plaice examined there were in all 13 supratemporal ossicles on the ocular side, as described by Traquair, but the recurrent portion of the canal, usually present at its anterior extremity in the adult (see figs. 23 and 29) was absent in this specimen.

The condition of this recurrent canal, it may be mentioned, varies considerably. On the eyeless side the rela tions of the supratemporal canal to the ossicles was essen tially the same. The sections revealed a curious absence of sense organs in the anterior portion of the supratemporal canal of the ocular side (fig. 23), there being only 4 sense organs as against 13 pores, but the canal on both sides of the body was not completely developed, like the infraorbital. On the eyeless side in the sections there was a break in the middle of the supratemporal canal occupied by three naked sense organs, only two of which, however, seemed to be canal organs. This break is of course converted into a canal later on in the ontogeny. The recurrent canal was also absent, and there were only eight pores as against 13 on the ocular side, although there were at least 7 sense organs as against 4, not counting the three naked sense organs above. The relative position of the sense organs was also different. In the lateral line itself the only difference of importance between the two sides was that the third or last otic sense organ of the ocular side was here innervated from the R. supratemporalis vagi, and hence belonged to the lateral canal, as in the Cod.

Infraorbital Canal (*inf. c.*).—After leaving the last supratemporal ossicle, in which this canal on the ocular side anastomoses with the lateral, it at once enters the pterotic, and immediately receives the hyomandibular canal, which comes up from below. There is no separate dermal pterotic as occasionally occurs in the Cod. The canal passing straight forwards, traverses the pterotic, sphenotic and right frontal. Arrived at the space between the first and second tuberosities (fig. 1) it anastomoses with the right supraorbital canal and turns sharply downwards almost at right angles, afterwards curving forwards

under the eye. After leaving the frontal the canal is protected only by the chain of very slender suborbital ossicles.
the posterior of which are sometimes called postorbitals,
and the last of which is attached to the frontal. These
ossicles are much more conspicuous in the Cod. In one
specimen examined there were 18 suborbitals on the ocular
side. On the eyeless side the suborbitals are both fewer
and larger, nor does the canal take such a wide sweep
forwards, and is hence shorter (cp. fig. 29). It leaves the
frontal in front almost exactly opposite the exit of the
right infraorbital. From this point the canal extends
forwards in a slight curve until it reaches the left
lachrymal to which the first suborbital ossicle is attached.
On the ocular side the lachrymal is quite distinct from
the suborbital chain, and in the specimen examined the
canal ended by a pore 11mm. behind and below the postero-
ventral extremity of the lachrymal. On the eyeless side
the canal passes on to the left lachrymal, on which it
terminates. There were 5 suborbital ossicles on this side,
but three of these were each divided into two. Traquair
describes 13 on the ocular side, and 8 on the eyeless. In
the sections the right infraorbital canal was not completely developed, the first 7 sense organs being yet
unenclosed in a canal, but lodged in small depressions of
the skin. The supratemporal and infraorbital canals are
therefore the last to be completed. The portion of the
canal in line with the " lateral line," and innervated by
the R. oticus facialis, contained 3 sense organs on the
ocular side, to the last of which attention may be drawn.
On the eyeless side anteriorly the canal courses downwards, but after leaving the lachrymal it turns sharply
backwards almost at right angles. The whole canal is
completely developed on this side, and is more robust.
Its anterior portion has fewer sense organs, but those 't

has are arranged more definitely with regard to the pores, *i.e.*, there is always one sense organ between two adjacent pores, except between pores 2 and 3, where a sense organ was absent. The part of the canal innervated by the R. oticus vii. also differs from that of the ocular side in that it has 3 pores, between the first two of which no sense organ was found, and only the first two sense organs were innervated by the R. oticus vii. The sense organ corresponding to the third or last otic was on this side innervated from the R. supratemporalis x., and hence belonged to the lateral canal, as in the Cod.

Hyomandibular canal (*hy. c.*).—This canal anastomoses behind on the pterotic with the infraorbital on the ocular side, and the lateral on the eyeless side, as above described. It passes forwards for a little and then turns sharply downwards on to the preoperculum. It courses downwards parallel with the anterior edge of the pre operculum, turns abruptly forwards with the same, and finally passes over the articular on to the dentary, on which it ends far in front by a pore. The positions of the sense organs and pores are shown in the chart. The canals of the two sides agree in all essential respects, except that in the sections, pore *ι* was found to be wanting. The full complement of sense organs, however, was present.

Right Supraorbital Canal (*sup. c.*).—This is the only complete supraorbital in the Plaice, the left being represented only by vestiges. It anastomoses with the right infraorbital between the first and second tuberosities on the frontal, and passes forwards for a short distance, and then gives off almost at right-angles the supraorbital commissure (*s. o. c.*), to join the supraorbital canal of the other side. It then passes forwards in the substance of the right frontal, leaves this in front and passes on to the right nasal, on which it ends by a pore. It has only 3 pores and 4 sense organs, omitting those in the commissure. It is curious to find that opposite the entry of the commissure there is a small blind diverticulum, containing no sense organs, and corresponding exactly to a similar structure found in the Cod (see fig. 23).*

The Left Supraorbital (*sup. c.'*).—Anastomoses with the left infraorbital just as on the opposite side, passes forwards within the left frontal for a short distance, and receives the commissure. Just opposite the latter point it gives off a surface pore below which may correspond to the blind sac of the ocular side. Between its anastomosis with the left infraorbital and reception of the commissure, it has one large sense organ† but no pore (cp. other side, fig. 23). In front of the commissure the canal leaves the frontal, but still lies in a depression on it, and ends blindly a short distance anterior to the commissure. This description enables us to compare the condition in the Plaice with that of the Turbot, as described by Traquair, and to correct the latter's figure of the Plaice in some slight particulars. The other supposed remnant of the left supraorbital canal, discovered by Traquair, is situated far in front on the ocular side of the body very near the dorsal edge and above but a little behind the anterior extremity of the right supraorbital (figs. 23 and 29, *sup. c.''*). It consists of a very small follicle situated, according to Traquair, on a minute ossicle representing a greatly reduced left nasal, and containing two surface pores, and, according to our sections, one sense organ.

---

* Cole, Trans. Linn. Soc., ser. ii., vol. vii., pp. 157 and 180.

† This sense organ, as well as the one on the left side of the supraorbital commissure, is innervated by the R. ophthalmicus superficialis vii. of the left side. This is quite a conspicuous nerve, in spite of the abortion of the greater part of its sensory canal. There can, therefore, be no question that these parts of the canal system belong to the supraorbital canal.

As, however, in the single series of sections on which fig. 23 was based it was innervated by the inner buccal of the right side, it either cannot represent the anterior end of the left supraorbital, or such an innervation must be anomalous. In the meantime, however, we shall follow Traquair, and regard it as belonging to the supraorbital system. Another very small follicle was found behind it, but it was very aborted and contained no sense organ. The innervation of the former structure suggests that it is a modified pit organ, especially as such occur typically in Teleosts at this region.

The Supraorbital commissure (*s. o. c.*).—Arises from the canal on the ocular side, as above described. A sense organ is situated just at its origin, which projects partly into the supraorbital canal itself. The commissure passes upwards and forwards almost at right-angles in the substance of the right frontal. It then bends sharply inwards at right angles (indicated by a circle in fig. 23), and at once enters the left frontal. Just at the turn an unpaired surface tubule (3) is given off, and this undoubtedly corresponds to the fourth unpaired median tubule of the Cod. Its position in the Plaice is only apparently anomalous. The commissure passes almost transversely but slightly forwards across the body in the left frontal. Arrived at the other side it bends gradually downwards but still forwards (a large sense organ being situated at the turn), and thus passes into the left supraorbital. The sense organ in the commissure on the left side is situated much higher up than on the other side, and is thus entirely within the commissure. It is innervated by the R. ophthalmicus superficialis vii. of the left side. The commissure also is passing forwards from right to left during the whole of its course.

To sum up the supraorbital system, the right supraorbital canal has 3 pores and 4 sense organs, the left supra-orbital 3 pores and 2 sense organs, and the supraorbital commissure one median pore and a pair of sense organs. Only the two extremities of the left supraorbital canal are present, the whole of the intermediate portion having aborted with the corresponding part of the left frontal. The first two pores and first sense organ of the left supra-orbital appear to correspond to the same on the right. The whole supraorbital system is supplied by the two RR. ophthalmici superficiales vii.

Besides the sense organs situated in the sensory canals, or canal organs, there are two other series of naked sense organs situated on the skin. These are known as Pit Organs and Terminal Buds. The former belong to the lateral line system, and are innervated by one or more of the four lateral line nerves. The latter are quite dis-tinct from the lateral line organs, and are innervated by an entirely separate system of nerves, the Ramus lateralis accessorius (= R. lateralis trigemini), belonging to the communis system. The terminal bud system has been reduced in the Plaice.

2.—The Nose (Fig 25).

The olfactory organ of both sides is described as if viewed from *above*, and not from the side. The external apertures of the nose, two on each side, are not symmetri-cally placed, and those of the left side are situated almost on the apparent mid-dorsal line of the body, immediately to the left of the anterior border of the left eye. The two apertures or nostrils of each side are sometimes called anterior and posterior nares, but the latter term is obviously inadmissible.

---

* For the olfactory organs of the Pleuronectidæ, see Kyle, Jour. Linn. Soc., xxvii., and 18th Ann. Rep. Fish. Board Scotland, 1899, Part iii.

Right Olfactory Organ.

Anterior Nostril (*a. nos.*).—A small aperture situated on the anterior aspect and almost at the apex of a largish tube.

Posterior Nostril (*p. nos.*).—A large aperture only slightly raised above the surface of the body.

Both nostrils open into a large nasal chamber (*n. ch.*), the external wall of which is perceptibly thickened, and the internal wall of which is thrown into a vertical series of large folds or olfactory laminæ (*o. lam.*). These laminæ bear the olfactory cells and sensory hairs, to which the olfactory nerve is distributed (*n. olf.*), and run in a longitudinal direction. The nasal chamber is therefore the sensory chamber of the nose, and in most fishes is the only one present. There were 9 olfactory laminæ in the right nasal chamber in our sections, but the most dorsal and ventral one is very small.

Into the nasal chamber open the nasal sacs, the walls of which are non-muscular and non-sensory, but contain innumerable goblet cells. They therefore have a secretory function at least. These sacs are as follows:—

Dorsal nasal sac (*n. sac.*[1]).—Opens into the nasal chamber behind and above. Passes forwards and bifurcates. The inner limb soon terminates, but the outer one passes far forwards, and although it is very narrow from side to side, it is very wide from above downwards. Its true extent therefore does not appear in the figure.

Antero-ventral nasal sac (*n. sac.*[2]).—Has a common opening with the postero-ventral sac into the nasal chamber behind and below. Its outer wall is continued backwards into that of the third sac. Passes far forwards, narrow from above downwards, but wide from side to side, between the mucous membrane of the mouth and the

palatine bone. In front a portion is sent upwards internally. at right-angles to the main portion and internally to
the palatine. This soon terminates, and the remainder of
the sac ends bluntly below the palatine.

Postero-ventral nasal sac (*n. sac.*[3]).—Opens into the
nasal chamber as above described. It is very irregular in
shape, and three-rayed in transverse section. It passes
far backwards just above the mucous membrane and to the
right of the palatine, being narrow behind and ending
blindly just over the mucous membrane of the mouth
below and internal to the sclerotic of the right eye.

Left Olfactory Organ.

The anterior and posterior nostrils (*a. nos.*[1], *p. nos.*[1])
are no different from those of the right side, except that
the anterior tube is smaller and the posterior nostril is
larger and more widely open.

The left nose is situated at a transverse level posterior
to that of the right, and is further much less developed.
This is most evident in the nasal chamber (*n. ch.*[1]), which
is obviously smaller and was in the sections only thrown
into 4 olfactory laminæ (*o. lam.*[1]), the dorsal one of these
being quite small and the ventral one smaller than the
two intermediate laminæ.* The figure does not admit of a
comparison as regards the nasal sacs, since their dimensions from above downwards cannot be shown.

There are only two nasal sacs on the left side as
follows :—

Dorsal nasal sac (*n. sac.*[4]).—Arises from the nasal
chamber dorsally from behind. At first its shape in transverse section is that of an inverted right-angle—the verti-

---

* The difference between the two olfactory organs is seen in the
diameter of the right and left olfactory nerves (fig. 25, *n. olf.*, *n. olf.*[1]).
The figure only illustrates the difference in diameter, the difference in
bulk is even greater.

cal limb lying between the nasal chamber and the inter-maxillary cartilage, and the horizontal limb passing inwards over the top of the cartilage. In front, however, the sac lies wholly on the top of the cartilage fitting over it in a curve like a cap. It ends very bluntly in front.

Ventral nasal sac (*n. sac.*[5]).—Its position in the figure is somewhat diagrammatic, as it is really situated under the nasal chamber and only partly under the dorsal sac. It arises from the chamber ventrally from behind, and also passes forwards. It is of very irregular shape, and gives off externally a small limb which soon terminates. Its true extent does not appear in the figure, as it is situated obliquely dorso-ventrally. It narrows very much in front and ends blindly near the outer skin, and between it and the intermaxillary cartilage.

There are no true posterior nares such as Kyle describes in one specimen of a *Cynoglossus*. The nasal sacs discharge their contents by the action of the jaw apparatus, and apparently fill again with sea water by the latter simply passing into the nasal chamber, and thence into the sacs, when the animal is swimming. No intrinsic muscular action is involved, nor are the nostrils valved.

The non-fixed parasitic Copepod *Bomolochus soleæ*, Claus, is not infrequently found in the nasal chamber of the Plaice.[*]

3.—The Eyes.

We have only space to consider those features in the structure and relations of the eyes and their accessory organs which are peculiar to Pleuronectid fishes, and are in some way associated with the asymmetry of the head. The eyes are situated very near the anterior limit of the

---

[*] For figures, see T. Scott, Eleventh Report Fish. Board Scotland, pl. v., figs. 1-10.

cranium, and the lower or right eye is slightly in front of the upper or left. The prominent interorbital ridge formed by the frontal is continued a little way round in front of each eye, that portion in front of the left being formed by the mesethmoid and left prefrontal, and that in front of the right eye by the right prefrontal. The interorbital ridge is continued backwards into the line of tuberosities. In the dead fish each eye lies in a little concavity, and the skin round it is loose and thrown into folds. This appearance may also be seen in the living fish, but it will be noticed that the eyes often project very markedly from the surface of the head, and that their axes may be parallel or even convergent, while after death they are widely divergent. There are no eyelids, and the cornea is flattened.

The orbits.—The left orbit (fig. 1) is bounded internally by the right frontal and an anterior process of the left frontal, anteriorly by the mesethmoid and left prefrontal, externally by the processes of the left frontal and prefrontal forming the pseudomesial ridge of Traquair, and posteriorly by the left frontal. When the cranium is placed dorsal surface uppermost the left orbit looks upwards and to the right. The right orbit is not bounded completely by bony walls as is the case with the left. Only internally and anteriorly do the bones adjacent to it lie close to the skin. Those are the right frontal and prefrontal. The bony structures external and posterior to the orbit are the parasphenoid bar and the alisphenoid. All these bones lie nearly in one plane, and the external and posterior walls of the orbit are formed by strong muscle masses. When the cranium is held dorsal surface uppermost the right orbit looks downwards and slightly laterally.

The interorbital septum is formed by the right

frontal and prefrontal above, and by the ethmoid cartilage below. As seen in the dried skull (fig. 3) these structures bound a large fenestra. A membranous sheet stretches across this fenestra from the ventral edge of the right frontal to the ethmoid cartilage and completes the septum. In front the ethmoid cartilage is perforated by a wide opening. This is the ethmoidal fenestra. It is seen in fig. 3, and through it the internal surface of the left prefrontal may be seen in the figure. When the cranium is held dorsal surface uppermost the interorbital septum is almost exactly horizontal.

The Recessus orbitalis.—This is the term applied by Holt[*] to an accessory of the organ of vision present in all Pleuronectid fishes. It is an evagination of the membranous wall of each orbit forming a sac which lies outside the orbit and the cavity of which communicates with that of the former by one or more openings. The recessus of the right eye lies immediately behind the bulb and just underneath the interorbital septum. To expose it the skin must be very carefully removed from the region immediately behind the eye from the interorbital ridge downwards. On removing a little connective tissue the organ is then seen. It is a conical sac of fatty appearance with the apex directed backwards. In a fish of about 22 inches in length it is about 1cm. in total length in the contracted condition. On cutting open its outer wall its cavity is seen to be somewhat reduced by bands of muscle fibres which cross it and are massed together on the internal wall. A seeker can be passed from its cavity into that of the orbit. On cutting away the skin round the eye and carefully removing the latter after dividing the optic nerve and eye muscles, the opening of the recessus can be seen immediately above the place where the

[*]Holt—Proc. Zool. Soc., No. 29, 1894, pp. 413-446.

external and inferior recti enter the orbit, and where the cavity of the latter is prolonged backwards for a little distance round the recti.

The recessus of the upper or left orbit has similar relations, but on account of the almost complete enclosure of the latter by bony structures it has a somewhat different position. It lies on the eyeless side of the head external to the fenestra bounded by the left frontal, left prefrontal, parasphenoid and alisphenoid. It must therefore be dissected from that side and is easily exposed by simply removing the skin and a little surrounding connective tissue. It is (in a preserved fish of 22 inches long) a round flattened sac of about 2cm. in diameter, of similar appearance and structure to the organ of the right side. It also opens into the orbit, and on removing the left eye the large aperture is easily seen immediately posterior and slightly above the place of origin of the inferior oblique muscle. It is situated beneath the pseudomesial ridge and pierces the soft wall of the fenestra mentioned.

We have stated that the eyeball can be protruded from the general surface of the head to a remarkable extent. Now while the eye muscles provide an effective apparatus for the retraction of the eye, there is apparently no muscular arrangement which can bring about protrusion. This appears to be the function of the recessus. In life the cavities of the orbit and that of the recessus with which the latter is in free communication are filled by fluid which apparently originates by an infiltration of lymph from the capillaries outside the orbital wall. The wall of the recessus being markedly muscular, it follows that its contraction will expel the contained fluid into the orbit and press on the internal surface of the eye-ball. Since the skin of the head round the eye is loose it yields, and the eye is accordingly protruded. Conversely the relaxation of the wall of the recessus and the contraction of the eye muscles effect the retraction of the eye-ball. If the eye in a fresh or living fish be gently pressed fluid passes into the recessus from the orbit and the former can be seen to swell.

The eye muscles (text-fig. 5) are the usual six pairs, two pairs of oblique muscles, and four pairs of straight muscles or recti.

The superior oblique muscles (*Obl. Sup.*).—Both right and left muscles take origin on the left side of the head. It will be remembered that the left prefrontal sends backwards a long process which fits into a groove on the dorsal surface of the parasphenoid. At the beginning of this process the prefrontal is deeply grooved for the reception of the ethmoid cartilage, the groove being formed by two horizontal bony laminæ, and on the upper of these laminæ and on the adjacent anterior concave surface of the pre frontal the oblique muscles take origin. The left superior muscle passes upwards and slightly backwards along the left side of the interorbital septum to its insertion on the eye-ball. The right muscle, however, immediately passes through the ethmoidal fenestra and so through the interorbital septum to the right orbit. It then passes upwards and backwards, diverging slightly from the left muscle, along the right side of the interorbital septum to its insertion in the right eye-ball.

The insertion of these muscles is peculiar. Each splits up near its distal extremity into anterior and posterior slips. The anterior slips (*obl. sup.*), which are the larger of the two, have wide insertions on the superior and anterior surfaces of the eye-balls. The posterior slips (*o. s. x.*) curve round the internal and superior surfaces of the eye-balls, crossing over the insertions of the superior

TEXT FIG. 5. Dissection of the Eye-Muscles. × ¾.

The Cranium is seen from the dorsal surface and slightly from the left side. The eyeballs have been pulled out from the orbits.

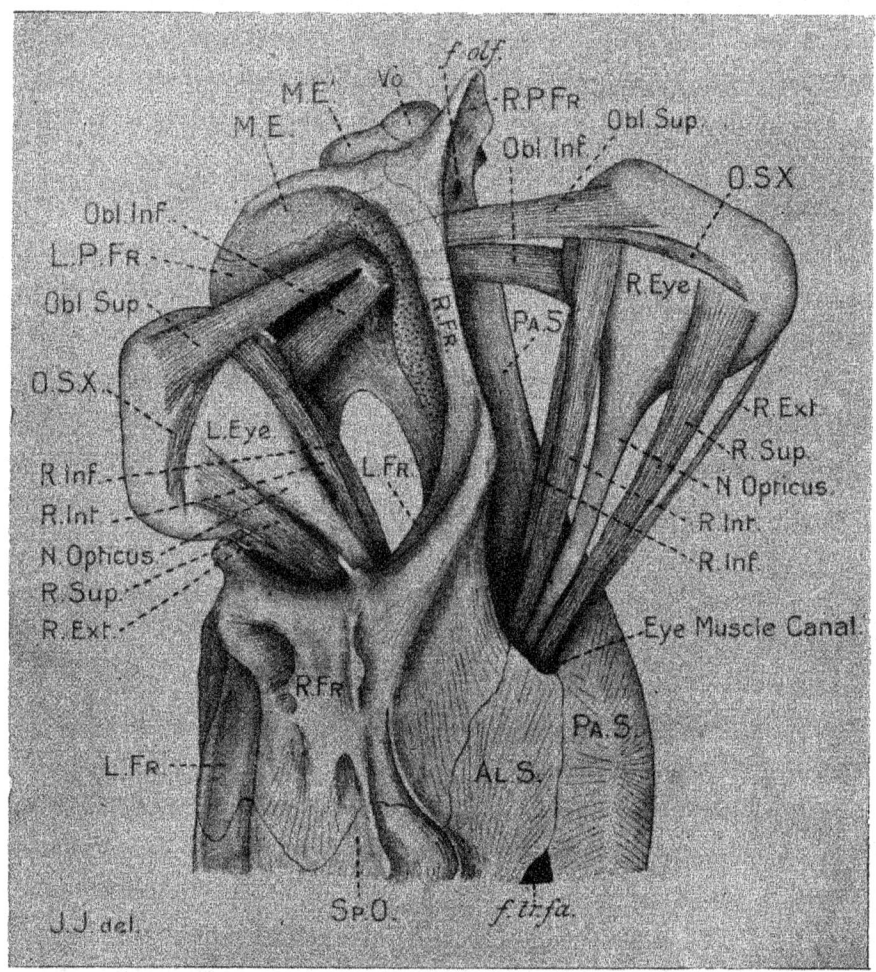

Obl. Sup., superior oblique; *Obl. Inf.*, inferior oblique; *O.S.X.*, rotatory slip of superior oblique; *R. Sup.*, superior rectus; *R. Inf.*, inferior rectus; *R. Int.*, internal rectus; *R. Ext.*, external rectus; *R.Fr.*, right frontal; *L.Fr.*, left frontal; *R.P.Fr.*, right prefrontal; *L.P.Fr.*, left prefrontal; *M.E.*, mesethmoid; *M.E¹.*, anterior portion of mesethmoid; *Vo.*, vomer; *Pa.S.*, parasphenoid; *Al.S.*, alisphenoid; *Sp.O.*, Sphenotic; *f.olf.*, olfactory foramen; *f.tr.fa.*, trigemino-facial foramen.

recti, and are inserted into the superior posterior surfaces of the eyes. These posterior slips are the rotatory slips of the superior oblique muscles, and their function* is to cause a rotation of the eye on its optical axis, a movement which is very exceptional among fishes as among higher vertebrata, and which is doubtless a special adaptation to the peculiar mode of life of the Pleuronectidæ. The extent of the rotation may be as much as one-eighth of a circle.

Inferior oblique muscles (*obl. inf.*).—Both right and left muscles originate in the left prefrontal in the same region as that from which the superior muscles take origin. The left muscle passes upwards and backwards along the external wall of the orbit and is inserted into the inferior middle surface of the eye-ball. The right muscle takes origin a little in front of the left, and, as in the case of the superior muscles, the proximal portions cross each other. It then passes through the ethmoidal fenestra with the superior muscle of its side, and passes upwards and backwards along the external wall of the right orbit, and is inserted in a corresponding position to that of the left eye. The inferior are slightly thicker than the superior muscles.

The superior recti (*r. sup.*).—These are strong muscle bundles originating in a strong oblique partition crossing the eye muscle canal. They run forwards in the latter, slowly diverging from each other, and emerge on either side of the interorbital septum. They are inserted into the superior and posterior margins of the eye-balls underneath the rotatory slips of the superior oblique muscles.

The inferior recti (*r. inf.*).—Also rather thick muscles which take origin far back in the eye-muscle canal at about the transverse level of the articulation of the head

* Bishop Harman—Journ. Anat. Phys., vol. xxxv., pp. 1-40, 1899.

of the hyomandibular, on the internal surface of the parasphenoid. They emerge from the canal into the deeper parts of the orbits and are inserted into the eye-balls on the inferior and middle surfaces, just underneath the insertion of the inferior obliques, so that their extremities are hidden by those of the latter muscles.

The internal recti (*r. int.*).—Not so strongly developed as either inferior or superior recti. They originate near the place of origin of the superior recti on the partition in the eye-muscle canal referred to. They run forwards in the orbit close to the interorbital septum, and are inserted into the eye-balls on the anterior and internal surface underneath the extremities of the superior obliques.

The external recti (*r. ext.*).—These are the most slender of all the eye muscles. They take origin on the internal surface of the parasphenoid far back in the eye-muscle canal, and leave the latter above and externally They are inserted into the external and posterior surfaces of the eye-ball. A large portion of the distal extremity of each is tendinous, and contains little contractile tissue.

With regard to the Bulbus oculi itself, we have only space to mention the more striking features in its anatomy.

Blood vessels.—The afferent vessels of the bulb are (1) the ophthalmic artery, the origin of which has been already described, and (2) a small vessel springing from the circulus cephalicus between the origins of the internal carotid arteries. The efferent vessel is the superior jugular vein which begins its course in the eye. These vessels lie in the eye-muscle canal. They emerge from the latter accompanied by the optic nerve of their side with which they are bound up by a common sheath of connective tissue. Arrived at the eye all three perforate the sclerotic. The ophthalmic artery breaks up in the choroid gland. The other artery appears to pass forwards with the optic nerve to the retina, but we are uncertain as to its precise distribution.

The sclerotic consists of two layers, an external layer of tough fibrous tissue into which the eye muscles are inserted, and an internal cartilaginous layer of some thickness. The cartilaginous sclerotic is perforated for the entrance of the optic nerve and the blood vessels, the fibrous layer becoming continuous with the connective tissue surrounding the latter structures. The cartilaginous sclerotic ceases at some distance from the pupil, and the fibrous layer with another layer which seems to be a continuation forwards of part of the choroid fuse with the skin of the head to form the cornea. In suitably prepared sections all these layers can be distinguished in the cornea, and the structure of the skin in that region does not differ materially from that in other parts of the head.

The Argentea.—Internal to the sclerotic and in close contact with its internal surface is the peculiar layer so named. It covers the whole internal surface of the sclerotic as far forwards as the iris. It has a white silvery appearance by reflected light, but is opaque to transmitted light. No structure beyond wavy bundles of very fine fibres can be made out in it. The silvery appearance is said to be due to minute crystals imbedded in a cellular tissue.

The Choroid.—This is the usual vascular and pigmented layer. It lies between the argentea and the retina, is closely adherent to the latter, and comes away with it when removed. Anteriorly it passes into the iris.

The Choroid Gland.—This structure is situated in the posterior wall of the bulb between the argentea and the choroid. It lies to the nasal side of the entrance of the optic nerve, is elongated in the longitudinal axis of the body, and curves round the nerve so that its concave margin faces the temporal side of the eye. It is not a gland *sensu stricto*, but a rete mirabile. The ophthalmic artery on entering the bulb immediately breaks up into a number of short capillaries which run transversely to the long axis of the gland; at the lateral margin these turn backwards on themselves, and open into a very prominent vein which traverses the whole length of the gland. From this vein a very short transverse trunk pierces the sclerotic and forms outside the bulb the distal extremity of the superior jugular vein. We are unaware of any plausible hypothesis as to the function of this organ.

The Retina presents no features of special interest. The radiation outwards of the fibres of the optic nerve towards its periphery is very striking. A prominent black line runs outwards from the place of entrance of the optic nerve to the temporal margin of the retina. This is the choroidal fissure, which here divides the retina, and by exposing the pigmented choroid beneath shews as a black line.

The Processus Falciformis and Campanula Halleri. — A delicate fold of the choroid projects through the choroidal fissure into the vitreous humour. This is the processus falciformis. Its distal extremity is swollen out into a bilobed pear-shaped enlargement—the Campanula Halleri, which is attached to the lens. These structures are said to form an accommodation apparatus. Accommodation in the fish eye is effected not by an alteration in the curvature of the lens but by its approximation to the retina through the contraction of the muscular tissue in the above campanula and processus. The elasticity of the suspensory ligament increases the distance between lens and retina.

Discussion of the Asymmetry

We have now ascertained the facts which justify a fuller discussion of the causes and course of the asymmetry of the head. To satisfactorily understand the considerations which we put forward, the reader will find it essential to follow our argument with dried crania both of a Plaice and Cod before him. We have already described the cranium in detail, but it will be useful to summarize those features relevant to the present discussion.

Note then that (A), the bony interorbital septum is formed by the right frontal alone, that it is a thin flattened bone lying in the morphological horizontal plane of the head, and that in front it is in contact with the ethmoid cartilage which is perforated by a wide fenestra; (B), that the left prefrontal has lost its primitive connection with the left frontal* but is still attached to the parasphenoid bar, while the right prefrontal, though still in contact with the right frontal has lost its connection with the parasphenoid bar, and is further distinctly anterior to the left bone; (C) that all the oblique muscles take origin from the left prefrontal.

Note now the differences between these relationships and those of the corresponding structures in the Cod's skull. In the latter (A), the fused frontals form a broad roof to the cranium, and internally and below two bony ridges form the bony portion of the interorbital septum, whilst the ethmoid cartilage is not perforated; (B), the right and left prefrontals are in contact with the frontals of their own side *and* with the parasphenoid bar; (C), the oblique muscles take origin from the upper portion of a strong membranous partition joining the frontal ridges

---

* The present connection between the left frontal and left prefrontal is purely secondary, as mentioned elsewhere.

with the ethmoidal cartilage, and practically they arise from the lower portions of the median frontal ridges.

Now if we suppose that the asymmetrical skull of the Plaice first began to appear in an ancestral form resembling the Cod, grave difficulties at once arise. For if a rotation from left to right of the orbital region of such a cranium took place, it is evident that although the left eye might ultimately look upwards, the right on the other hand must also move and would tend to become buried in the tissues of the head. Moreover it is probable that if a round fish such as the Cod adopted sedentary habits on the sea bottom, the flattening, if it occurred at all, would be a dorso-ventral one, as in the case of the skate.

But if we assume that a laterally compressed fish (like *Zeus*, for instance) took to a bottom habit, and began to lie on one side of its body, the changes necessary to bring about such asymmetry as we find in the Plaice would be much less violent. In such a form we may suppose that, following the lateral compression of the body, the eyes would have moved to near the dorsal edge of the head, whilst the frontals would have become greatly compressed from side to side and probably elongated dorso-ventrally, like the right frontal of the Plaice. The eyes therefore being now close together near the dorsal median line of the head, require to travel so much the lesser distance.

Now when such a fish assumed a bottom living habit, lying on (say) the left side of its body, any variations in the position of the left eye bringing it nearer the middle line than the right (*i.e.*, nearer the upper side) would be of great advantage. This approximation to the middle line would be attained either by the attenuation of the left frontal or by a shifting of both frontals towards the right side. In the skull of the Plaice both these things have happened.

It is very remarkable that comparatively slight changes would have sufficed to bring about the distortion of the Plaice's skull starting with such conditions and tendencies as we have indicated above. In the head of the Plaice the orbital region only has suffered extensive change. The otic region has undergone practically no change, and the prefrontal region certainly less than the orbital. The jaw apparatus is not affected at all by the torsion of the orbit, but has an asymmetry of its own (as described above), and this latter doubtless explains the asymmetry of the prefrontal region, which is of a different character to that of the orbital. We have, therefore, now to consider whether such relationships in the anatomy of the skull and eyes as obtain in the Plaice can be reasonably expected to have followed from such a course of evolution as we have sketched above. There appear to us to be four difficulties requiring explanation. These are as follows :—

(1.) The relations of the two prefrontals. As the right frontal bent over further to the right side, the right eye would be forced downwards. To make room for it the parasphenoid bar most probably bent over to the left to the position in which we now actually find it. The prefrontal region lagged behind in this shifting, being unaffected except as regards the movements of the adjacent parts (orbital region and jaw apparatus). If a Cod's skull be examined it will be seen that the abortion of the left frontal would result in the separation of the left prefrontal from the remainder of the frontal area, and further that the greater rotation of the frontal over the prefrontal area would result *either* in the separation of the right prefrontal from the right frontal, *or* in its separation from the parasphenoid bar. Now the latter having itself rotated to the left, we therefore find that in the Plaice the right prefrontal retained its connection with the right frontal, but lost its connection with the parasphenoid bar. *After* the rotation of the orbital region the left prefrontal grew backwards and the left frontal forwards, so as to form the secondary junction of these two bones already referred to.

(2.) The downward shifting of the oblique eye muscles from practically the frontal to the parasphenoid bar. This possibly began in the symmetrical but laterally flattened ancestor, as the frontals became attenuated from side to side and deepened dorso-ventrally. But it was continued in the asymmetrical fish. The object of the shifting of the eyes was to make dorsal vision possible with both eyes. In the Cod the origins of the oblique muscles are best adapted for lateral vision and for visual axes in approximately the same straight line. In the Plaice, however, the visual axes are directed dorsally, and to bring this about the oblique muscles had to shift ventrally so as to exercise a more effective pull over the eyes. The origins of these muscles therefore moved downwards towards the ventral region of the cranium and ultimately to the parasphenoid bar.

(3.) The attachment of all the oblique muscles to the left prefrontal. We have already pointed out that the right prefrontal has lost its connection with the parasphenoid bar. This would doubtless have occurred in the earlier stages of the rotation of the eyes. When, therefore, in the final stages of the torsion, the oblique muscles in their downward passage over the interorbital septum arrived at the bar, the right prefrontal must have already left it. They thereupon continued their migration on to the left side and became attached to the left prefrontal—the only convenient attachment remaining.

(4.) The passage of the right oblique muscles through

the ethmoid fenestra. Primitively these muscles did *not* pass through a fenestra. In the Cod's cranium this does not exist, but there is a cartilaginous wall in *front* of the origin of the oblique muscles. In the migration ventrally of the origins of these muscles the latter passed behind and across to the left of the ethmoid cartilage to reach the left prefrontal. The former then grew up behind and over the muscles, thus forming the fenestra.

We have not studied the anatomy of the head in other Pleuronectids, and are unable to say whether the relations of the eye muscles above described are general. Cunningham (*op. cit.*) has described those relations in the sole, and it appears that they differ considerably from those we find in the Plaice. In the sole " the superior oblique of the ventral [right] eye arises from the small left [right is evidently meant] ectethmoid which is on the right edge of the interorbital septum; the inferior oblique arises from the external surface of the parasphenoid below the right ectethmoid. But both oblique muscles of the left or dorsal eye arise from the inner surface of the left ectethmoid." Owing to this disposition the direction of the oblique muscles of the left eye is at right-angles to that of those of the right, and this difference has resulted from a rotation of the left ectethmoid.

The asymmetry of the sole was produced according to Cunningham by the constant contraction of the oblique muscles of the left eye, so as to ' turn the pupil into a horizontal direction and look along the edge of the head." The eye thus pressed on the interorbital septum, and led to the absorption and distortion of the latter. At the same time the fulcrum of this pull (the left ectethmoid) has itself undergone considerable rotation " so that the surface of attachment [of the oblique muscles] which originally looked outwards to the left came to look upwards."

In the absence of figures we find it difficult to follow Cunningham's explanation, and we also find it difficult to believe that the same mechanical strain—that of the oblique muscles between the left ectethmoid and the left optic bulb—could have, at the same time, (1) rotated the bulb, (2) pressed on the interorbital septum so as to bend that over to the right, and (3) rotated the ectethmoid through a right angle. And we regard it as unjustifiable to base such a mechanical explanation on the attachments of the muscles *in an already asymmetrical skull.* It is inconceivable that the oblique muscles were attached in the immediate symmetrical ancestor of the sole in the same way that they are now. In the Cod, for example, we find them arising symmetrically from the interorbital septum, and therefore any discussion of the possibility of those muscles producing distortion of the head should be based on the conditions present in an unmodified symmetrical cranium.

But whatever the conditions are in the sole we find that in the Plaice Cunningham's hypothesis is an impossible one. Unlike the sole, all the oblique muscles are attached to the left prefrontal (= Cunningham's left ectethmoid), and the latter we regard as the least altered element in the orbital or preorbital regions. That the surface to which the oblique muscles are attached now looks upwards we regard as more simply explained by supposing that the upper portion of the left prefrontal like most of the left frontal, with which it was most probably suturally attached, suffered abortion in the shifting of the left eye. And the shifting of the origins of the oblique muscles to it has been a result of, or has been concomitantly brought about by, the approximation of the eyes, and the increasing tendency to dorsal vision.

4.—The Ear (Fig. 24).

The auditory nerve is described with the other cranial nerves.

As in Teleosts generally the ear is only externally enclosed by the ear ossicles. Hence it projects freely into the skull cavity internally, and is only separated from the brain by its own and the brain membranes.

The **utricular sac** consists of a central chamber (*utr.*, often divided into three portions, two of which constitute the utriculus *sensu stricto* and the third the posterior utricular sinus), and a wide ventral chamber or superior utricular sinus or canal commissure (*utr.*[1]). These, however, may conveniently be called the central and vertical chambers of the utriculus. From the anterior end of the central chamber connections are effected with the ampullæ of the anterior and external semicircular canals, at the posterior end with the ampulla of the posterior semicircular canal and the other extremity of the external canal. The vertical chamber rises up almost at right-angles from about the centre of the central chamber, and receives above the upper extremities of the anterior and posterior semicircular canals. There is only one sense organ in the utriculus, the macula acustica recessus utriculi (*m. r. u.*), situated in a slight depression of the central chamber in front (Recessus utriculi). The three semicircular canals are disposed as follows :—

Canalis anterior (*a. s. c.*).—Just above its connection with the central chamber of the utriculus it swells into a large ampulla anterior (*amp. a.*), the outer wall of which bears a transversely extending sense organ and ridge, the crista acustica ampullæ anterioris (*ant. cr.*). Above, the canal passes upwards and backwards into the vertical utricular chamber.

Canalis externus (*e. s. c.*).—Sometimes called the horizontal semicircular canal to distinguish it from the other two vertical canals. Swells into the ampulla externa (*amp. e.*) immediately behind its connection with the utriculus in front. The crista acustica ampullæ externæ (*ext. cr.*) is situated on the floor of the ampulla behind, but in front it ascends upwards and outwards, so as to invade its external wall. The canal passes hori zontally downwards and backwards, external to the rest of the ear, to communicate with the utriculus behind.

Canalis posterior (*p. s. c.*).—The large ampulla posterior (*amp. p.*) occupies a similar position to that of the anterior canal. Its outer wall contains a transverse sense organ and ridge, the crista acustica ampullæ posterioris (*post. cr.*). Above, the canal passes upwards and forwards into the vertical utricular chamber.

The **Sacculus** (*sac.*) appears at first to be an absolutely closed bag in the Plaice, closely opposed to the utriculus below, but having no connection with it. From its inner wall it sends upwards, however, a blind finger shaped process which must represent a vestigial ductus endolymphaticus (*d. end.*). There is no saccus endolymphaticus. Into the base of the ductus there open two very minute capillary tubes (see figure), and as each arises from, and communicates with, the utriculus, they must together represent the canalis utriculo-saccularis. There is thus only a very slight communication between the sacculus and utriculus, and this of a very curious nature.

The sacculus contains the large hard saccular otolith deposited in concentric laminæ (*oto.*), and often called *the* otolith.* There is, however, a small but quite conspicuous

---

* See Reibisch, Wiss. Meeresuntersuch, Abth. Kiel, N.F., Bd. iv., p. 233 for an interesting discussion of the relation of the otolith to the age of the fish. Reibisch finds that it is possible to deduce the age of a Plaice from the conformations of the otolith.

otolith in connection with the macula acustica recessus utriculi, and another with the papilla acustica lagenæ, but none were observed in relation to the three cristæ acusticæ, although it is possible that small concretions of chalk were there also. There is a very large sense organ on the inner wall of the sacculus in front, known as the macula acustica sacculi (*m. a. s.*), and behind, the sacculus is prolonged upwards and backwards into the digitiform lagena (*lag.*), otherwise known as the recessus sacculi or cochlea, the inner wall of which, at about the roof, contains another sense organ, the papilla acustica lagenæ (*p. a. l.*). A constriction or definite sacculo-cochlear canal is not differentiated.

The sense organ on the floor of the utriculus known as the macula acustica neglecta is entirely wanting in the Plaice.

The ear and auditory nerve of *Pleuronectes flesus* have been described by G. Retzius,† who, with other authors, refers to the utricular, saccular and lagenar otoliths respectively as the lapillus, sagitta, and asteriscus. His description agrees very closely with ours of the Plaice, so that it is only necessary to point out that he did not succeed in finding either a reduced ductus endolymphaticus or a utriculo-saccular connection. With regard to the latter he says:—"An der unteren Wand des Utriculus, ungefähr gerade unter dem Sinus superior, findet sich eine kleine, trichterförmige, hohle Verlängerung des Utriculus, welche nach hinten geht und sich zu einer feinen Düte verschmälernd an die Wand des Utriculus legt; ob aber dieser Canal blind endigt, was ich am meisten glaube, oder ob er vielleicht in den Sacculus (als ein Canalis communicans) ausmündet, kann ich nicht mit Bestimmtheit angeben."

† Anat. Unters., 1872, p. 52, Taf. v., figs. 10—18.

G.—THE REPRODUCTIVE ORGANS.

Some difficulty will be experienced in determining the anatomy of the reproductive organs, on account of the fact that sexual maturity does not occur until the Plaice has grown to a relatively large size, and that to examine the varying relations of the system in a really satisfactory manner fishes in various phases of reproductive activity must be examined. The ordinary marketable Plaice is as a general rule an immature fish. There is some consider able difference in the size and age at which it becomes sexually mature in different localities, but it will be suffi cient for our present purpose to say that under 17 inches of total body length in the female, and 14 inches in the male, the reproductive organs are generally immature. Only fishes of those sizes should therefore be dissected for these organs, and they are best examined some little time prior to the spawning season, that is during November to January.

We may here define several terms used in the following pages. The Plaice is spoken of as " mature " when it first begins to produce eggs capable of fertilization, or, in the male, functional spermatozoa. It is " ripe " when the ovary becomes distended with mature ova, that is immediately before the spawning season. It is " spent " or " shotten " when all the ova have been extruded in the act of spawning. The same terms with the same signifi cance are applied to the various phases of the male in respect to the conditions of the testes.

Considerable differences in the condition of the reproductive organs may then be expected in Plaice of different sizes or of mature fishes captured at different times in the year. In the female fish of about $1\frac{1}{2}$ inches long ovaries and ovarian eggs are certainly developed, but the organs are represented by very minute structures situated in the

o

posterior wall of the body cavity. At this age, and for some considerable time afterwards, it is quite impossible to distinguish between male and female gonads without a microscopical examination. The growth of ovaries or testes is extremely slow during the first two years of life, that is up to 6-8 inches long. After this age the ovary can be readily distinguished from the testis. It is a paired conical shaped structure. The base just projects into the posterior part of the body cavity; the apex projects backwards towards the tail, lying between the haemal spines and the muscles of the trunk. The oviducts are extremely difficult to dissect and do not open externally. In the following year the organ rapidly develops. In male Plaice of the same size the testes are paired ridges of the posterior body wall projecting forwards into the body cavity, but having no extension backwards as in the case of the ovaries.

The **Female organs.**—Fig. 20 represents the condition of the ovary in a mature and nearly ripe Plaice. The specimen figured was captured in December. The ovary is seen to project far forward into the body cavity, displacing various organs from their normal positions. Part of the posterior extension of the organ is indicated in the figure; it really extends backwards to near the root of the tail. Its posterior portion lies along the external surface of the haemal spines, and is only separated from these by loose areolar tissue. The overlying muscles of the trunk are very thin, and occasionally the ovary appears to be covered only by integument and connective tissue. In this condition it can be felt externally as a hard pad lying on either side of the ventral portion of the body behind the anus. The ovary of the ocular side appears to be, sometimes at least, more strongly developed than that of the eyeless side.

On spawning a very marked change takes place. Fig. 21 represents the condition of a mature and spent female. The ovary is seen to be considerably retracted, and is only just visible on opening the body cavity. Its walls are soft and flaccid, and enclose a large cavity. The posterior extension still exists, and is indeed not much shorter than in the ripe specimen. The condition of the fish is indicated externally by a shallow groove running backwards on either side in the position formerly occupied by the pad spoken of above. The flesh is lean, and the fish is generally regarded as in poor condition as a food.

It appears* that after spawning the ovary never reverts to its former immature condition. That is, it is always possible to distinguish between a spent mature, and an immature fish. Similar contrasts are exhibited by the various phases of the testes, but on account of the relatively small volume of these organs the changes are not so striking and afford no external indications.

The ovary of the Plaice, like that of the majority of Teleostean fishes is a sac the wall of which is continuous with that of the oviduct. This is the cystoarian condition, and for an understanding of the morphology of such an ovary the other common type met with among Teleostei, the elasmoarian ovary, must be studied.† The internal wall of the cystoarian ovary corresponds to the external face of the peritoneal lamella which forms the elasmoarian organ, and to the outer visible surface of the ovary of an Elasmobranch fish—the surface from which the ova dehisce into the general body cavity. There can be no doubt from the work of Balfour and Parker‡ on *Lepidosteus*, that the cavity of the cystoarian ovary into which

* See Holt, Journ. Mar. Biol. Ass., vol. ii., p. 363.

† See Howes—Hermaphrodite genitalia of the codfish. Jour. Linn. Soc., London, vol. xxiii., 1891, pp. 539-557.

‡ Structure and development of *Lepidosteus*, Phil. Trans., vol. clxxiii., Part 2, 1882.

the ova dehisce is a closed portion of the cœlom. The cavity of the oviduct is also a portion of the cœlom. It does not appear to be a Müllerian duct like the oviduct of all other vertebrata, and its morphology is exceedingly obscure. It has no homologue among the genital structures of the male fish. The external opening of the oviduct (*Od.*, fig. 20, pl. V.) lies just within the lip of the anus on the posterior side of the latter. It may appear to be outside the anus on the ventral body wall, and its variable situation is due to the extent to which the terminal portion of the rectum is contracted. In the immature fish the opening does not apparently exist, and even in the mature but unripe individual we have been unable to satisfy ourselves of its existence If in a ripe female Plaice the body over the region of the ovaries be gently pressed, the mature eggs issue in a thick stream, and the opening can then be easily seen as a transverse slit behind the anus. After extrusion of all the eggs this slit seems to close up by the adhesion of its edges, and in dissecting such a spent specimen, though the terminal part of the oviduct can be traced to just behind the anus, some little pressure with a blunt seeker is required to force a passage. In the nearly ripe fish which has not yet spawned even a moderate pressure on the abdomen may not cause the extrusion of any eggs, though dissection may shew that mature ova are present in considerable number lying loose in the cavity of the ovary. The more nearly ripe the fish is, the more easily are the eggs expelled, until in some cases merely lifting it from the water in which it is lying may cause the eggs to run from it. There is generally no bleeding from the edges of the forced opening in a nearly ripe fish, and sections of the region of the body wall behind the anus shew that it is formed almost entirely of dense fibrous connective tissue without many blood vessels. It appears to be the case that the efferent portion of the oviduct opens before each spawning, and closes again by adhesion of its walls when the act of spawning is over.

The external oviducal opening leads into a short chamber ($Od.'$) into which both right and left ovaries open. $Od.''$ is the cavity of the right organ. The septum in the figure is the fused internal walls of both ovaries. All this terminal chamber in the ripe fish is filled up with ova which have dehisced from the ovigerous lamellæ. On the internal walls of the ovary are the ovigerous lamellæ, longitudinal folds of the wall in which the ova are developed. Text-fig. 3 represents part of a transverse section of the ovarian wall in a spent fish and shews a single ovigerous lamella. In this condition the wall is thick. Externally there is a loose connective tissue layer, and in the thickness of this a thin sheet of black pigment. Internal to this layer is an investment of unstriated muscle fibres of some thickness in the spent ovary, but very thin in the ripe condition. Within this, and filling up the thickness of the lamella, is a somewhat dense connective tissue stroma. The surface of the lamella is formed by an epithelium which in places has a rather obscure structure, but here and there contains patches of small rounded cells, obviously a germinal epithelium. From this the ova proliferate into the thickness of the lamella, and come to lie freely in the stroma, at first near the surface of the former. Many such ova of different sizes are represented in the section. The largest have a distinct zona radiata, the nucleus is large with a distinct nuclear membrane and with diffused chromatin. An obvious ring of large spherical nucleoli is seen in contact with the nuclear membrane. Often the nucleus is con tracted away from the membrane, leaving a clear space.

The **Male organs**.—The testes are undivided flattened

masses lying one on each side of the 1st axonost on the posterior wall of the body cavity. They are indistinctly lobulated, but not follicular as in the Cod and so many other Teleosts. They are much smaller than the ovaries in the female and project forwards only a little way into the body cavity. There is no posterior extension beyond the limits of the latter at any phase in the spawning cycle. They are tubular glands, but the walls of the seminiferous tubules are closely opposed and there is little or no connective tissue between them. The whole organ appears therefore as if divided up by a system of irregular trabeculæ in the meshes of which are massed the spermatozoa.

Towards the ventral and anterior extremity of the testes the volume of the organ rapidly diminishes, and if traced in serial sections the number of tubules becomes much less. Finally three or four such run forwards on each side laterally and dorsally to the terminal portion of the ureter. These unite, and a single duct on each side runs alongside the latter and opens into its terminal portion quite near the urinary papilla. The terminal portion of the ureter is therefore a urinogenital sinus. If the abdomen of a male fish when near the spawning season be gently pressed, the seminal fluid issues from this papilla. In the female it is a urinary papilla only, in the male a urinogenital papilla.

APPENDIX—ECONOMIC.

A.—LIFE-HISTORY AND HABITS.

Spawning.—About the beginning of the year the reproductive organs of mature Plaice become ripe and spawning commences. Spawning—that is the complete extrusion of the contained ripe ova and spermatozoa, lasts over a considerable period on any one fishing ground. This is due to the fact that it requires some time for any one fish to extrude all its ova, and also to the considerable variation in the time of ripening of the reproductive organs among all the mature fish present on the spawning ground. The duration of this " spawning season " varies in the seas round the British Islands. In the Danish seas it begins in November, attains a maximum in January and February, and ends in April. In the North Sea on the Scottish side it lasts from the middle of January till the end of May, with the maximum at the beginning of March. In Loch Eyne in the Firth of Clyde in 1898 no Plaice eggs were found till the middle of February and none after June; the maximum number was found in April. In the Irish Sea the exact limits of the season have not been determined, but certainly it begins later than in the North Sea. The maximum period as determined by the abundance of ripe female fish is at the end of March.

From one quarter to half a million eggs are extruded by a single female Plaice during its spawning season, the average number on the various fishing grounds being about 300,000. This number of eggs is small when compared with that of many other flat and round fishes. Generally the larger the fish the greater the number of eggs yielded. It is evident that during the spawning season an immense number of eggs must be present in the seas round the spawning grounds, and this has been approximately determined in various places. In a series of remarkable researches on the quantitative estimation of planktonic organisms Hensen* has made an approximate determination of the floating fish eggs in various seas. In the North Sea in 1895 he found during three voyages an average number of 56·8209 Plaice eggs per square metre of the track of the vessel. From the known area of the North Sea (547,623 millions of square metres) Hensen determined the total number of Plaice eggs in that area to be 31·117 billions. It is evident that this number can only be approximately accurate, but there are reasons for believing that it really underestimates the total quantity. In much the same way, Williamson† determined the total quantity of Plaice eggs present in Loch Fyne during the first eight months of 1898, and found a total number of 483 millions to be present in the loch during that time.

The **Egg** is one of the largest of those belonging to Pleuronectid fishes, and is on that account easily recognised in plankton collections. Its transverse diameter varies from 1·63 to 2·11 mm. It is enclosed in a fairly tough capsule, the outer surface of which is finely corrugated; at one place, beneath which the germinal disc forms, there is a minute opening in the capsule—the micropyle. There is no oil globule, and the contents of the egg are of a glassy transparency and apparently homogeneous. A small perivitelline space is present between the yolk mass and the capsule.

Before ripening the ovary of the Plaice contains only opaque eggs considerably smaller than those extruded

---

* Hensen u. Apstein—Wiss. Meeresuntersuch., Kiel Commission, Bd. 2 (N. F.) Heft 2, p. 71, 1897.

† 17th An. Report Scottish Fish. Bd., p. 79, 1898.

after ripening. In the final stage of maturation before spawning occurs these small opaque eggs acquire the characters of the ripe pelagic egg.* The nucleus breaks down and the chromatic matter becomes rearranged, and at the same time fluid of a low specific gravity, secreted by the follicular epithelium enters. As a result of this imbibition of fluid the yolk becomes altered in such a way as to become nearly perfectly transparent. At the same time the egg becomes larger and its specific gravity becomes less. The immature ovarian egg has a mean diameter of 1·21mm. and a mean volume of 0·9276 cub. mm. It is heavier than sea water. The mature egg has a mean diameter of 1·88mm. and a mean volume of 3·479 cub. mm. It is very slightly lighter than normal sea water. The change in specific gravity during maturation is from about 1·07 in the immature to about 1·025 in the mature egg. Changes of this nature are general in the maturation of nearly all Teleostean food fishes, and as a result the eggs are pelagic—that is they float freely near the surface of the sea when extruded. In the eggs belonging to the other type—the demersal egg, of which the egg of the herring is the most familiar example—the specific gravity is greater than that of normal sea-water. As a result of this the eggs undergo their development lying on the sea bottom.

The Plaice takes about two weeks to extrude the whole contents of its ovary. Obviously in the limited space at the disposal of that organ ripening of all the eggs present would be impossible without injury to the fish. Spawning is therefore intermittent during the season of the fish, only comparatively few eggs being discharged at one time. As the latter mature they dehisce from the walls of the ovigerous lamellæ and accumulate

* Fulton—16th An. Report Scottish Fish. Bd., Pt. iii., p. 88, 1897.

in the cavities of ovaries and oviducts, and are expelled at intervals until the whole organs are exhausted. The fish is then said to be " spent."

Fertilization takes place in the surrounding sea water into which the eggs are extruded. During the spawning season males and females are crowded together on the spawning grounds, and the spermatozoa of the former are shed simultaneously with the eggs. It is impossible to say what proportion of the eggs extruded escape fertilization. Probably it is very small, since unfertilized eggs are not often seen in plankton collections. Fertilized and unfertilized eggs would both rise towards the surface, but in the course of a few days the unfertilized egg becomes opaque, heavier, and finally sinks to the bottom and decomposes.

Embryonic Development.—The period occupied by embryonic development varies with the temperature of the water in which the eggs are contained. The lower the temperature the longer is the period between fertilization and hatching. H. Dannevig* found that at 5·2°C. 21 days were required; at 6°, 18¼ days; at 10° 12 days; and at 12° 10½ days. Among Pleuronectidæ there is a general correspondence between the size of the egg and the developmental period. Thus the Flounder (*Pl. flesus*) with an egg of 0·95mm. in diameter hatches out in 4½ days at 10°C. At the same temperature the Sole (*Solea vulgaris*), which has an egg of 1·4mm. in diameter, requires 10, and the Plaice, with an egg of 1·8mm., 12 days.

The changes in the ovum immediately succeeding fertilization have not been closely studied by any author. The spermatozoon enters the egg through the micropyle; at this time segregation of the protoplasm takes place, and the latter which had previously been diffused round

* 13th An. Report Scottish Fish. Bd., Pt. iii., p. 147, 1894.

the periphery of the yolk mass now collects into a lenticular mass—the germinal disc. When the egg is floating freely, the germinal disc lies downwards, the yolk being uppermost. This lower part of the egg has been termed the "animal pole," the upper part the "vegetative pole." A few hours after fertilization meroblastic segmentation begins, by the formation of a single vertical furrow which divides the germinal disc into two blastomeres. This is followed soon by a second furrow transverse to the first and four equal blastomeres are formed. Up to the 8-celled stage at least there are no horizontal segmentation planes. After this stage both vertical and horizontal division planes occur and the germinal disc segments with increasing irregularity until it becomes a many-layered mass of small cells. The blastoderm is thus formed.

The rim of the blastoderm now begins to grow outwards and to envelop the yolk mass by a process of epibolic gastrulation. Towards the end of the 2nd day[†] it has extended over the yolk so as to cover about 70° of the latter when seen in optical section. The growing margin of the blastoderm is slightly thickened. The space enclosed within this blastodermic ring where the yolk mass comes to the surface is the blastopore, and with the growth of the blastoderm past the equator of the ovum its area continually diminishes. At first it is circular in form and then becomes oval. After 6-7 days from fertilization it disappears entirely.

The embryo begins to be raised off on the 3rd day after fertilization. It lies in such a position that the extremity which becomes the posterior one is situated against the edge of the blastopore. On the 4th day it has lengthened out considerably. The notochord and neural

[†] Development is supposed to be effected at a temperature of about 7° C.

axis are laid down, and the main divisions of the brain are visible. The optic vesicles are formed but are not yet invaginated, and no mesoblastic somites are as yet discernible. The tail projects as a slight swelling into the blastoporic area. Underneath the tail a small vesicle has appeared, which increases in size with the waning of the blastopore. This is Kupffer's vesicle (fig. 32), a structure characteristic of the Teleostean embryo. It lies immediately beneath the posterior extremity of the notochord, and in close relation to a slight depression under the thickened blastodermic rim beneath the tail protuberance —the "prostoma." Its cavity is bounded by a regular layer of hypoblastic cells, or its dorsal wall is so formed, and its ventral wall is formed by the yolk. In some Teleosts it communicates with the exterior by a narrow opening. It is the invagination cavity of the (masked) Teleostean gastrula, and represents the archenteron. But from its large size and persistence it is probably a functional larval organ, and Sumner* has suggested that it subserves the nutrition of the embryo by aiding in the absorption of the yolk.

On the 5th day constriction of the optic vesicles from the mid-brain is complete, and on the 6th they are invaginated and the lenses are formed so as to lie in the openings of the optic cups. About 18 pairs of somites are now present. The trunk has elongated considerably and the head and tail are now constricted off from the yolk mass; 8 days after fertilization the auditory vesicles are present, the heart is formed and is beating, and the vitelline circulation is being laid down. The tail has increased considerably in length and about 30 somites are present. On the 11th day rudiments of the pectoral fin folds are present.

* Mem. New York Acad., vol. ii, 1900, pp. 47-83.

Pigmentation of the embryo begins on the 9th day by the formation of a row of yellow branching chromatophores on either side of the body; on succeeding days these become very abundant and extend on to the head and cover uniformly the trunk and tail. Black pigment appears on the 13th day as a row of round chromatophores on each side of the body. Later on these become abundant and of a branching form. On the 14th day the eye becomes pigmented, and has a greenish-golden sheen.

The little fish hatches out from the egg on the 17th day. It (fig. 35) is about 6·5mm. in total length—a relatively large size among newly hatched Pleuronectids. The yolk sac is very large. A continuous broad fin runs along dorsal and ventral margins of the body and round the tail. The notochord is straight at the tip, and only rudiments of the pectoral fins are present. It is covered (except the yolk sac) with bright canary-yellow chromatophores, and branching black chromatophores are situated along either side of the body. The eyes are greenish-gold in colour. The mouth is open; the gut is a nearly straight tube slightly dilated at one part and terminating in the anus at the posterior margin of the yolk sac. The œsophagus is open, but has an exceedingly contracted lumen, and there is an extensive yolk sac circulation. A pronephros as described above also exists. The urocyst is in connection with the hind gut, and does not open directly to the exterior. The young fish is still perfectly symmetrical.

The larval period and metamorphosis.—The larval period lasts from the time of hatching until the definitive form and asymmetry of the adult has been acquired; that is till about 6 weeks from hatching. For the first week there is little change in the larva except that during that time the yolk is being gradually absorbed, and at the end of 8 days the yolk sac has disappeared. The larva now begins to feed. It is probable that it feeds before the yolk has been entirely absorbed. The food is necessarily small, consisting of diatoms and larval molluscs. It grows very slowly, and at the age of 25 days from hatching is only about 7·8mm. in total length and about 1·6mm. in height. After this the increase in height is relatively greater than that in length. Larval crustacea now form the principal food. The tail, which has hitherto preserved its embryonic homocercal character, now bends upwards at the tip, and ventral to this upturned portion fin rays begin to be formed (fig. 36).

Up to 30 days after hatching the larva has retained its bilateral symmetry, and at this stage is precisely similar in form to the ordinary symmetrical Teleostean larva. The greater relative growth dorso-ventrally than longitudinally indicates the beginning of the metamorphosis, and after the 30th day the left eye begins to move dorsally and anteriorly. 40 days after hatching it appears on the dorsal margin of the head just anterior to the right eye. On the 45th day the left eye has attained its definitive position on the apparent right side dorsal and anterior to the right eye. The larva is about $13\frac{1}{2}$mm. long and $6\frac{1}{2}$mm. high. During the period in which the eyes are rotating the young fish gradually acquires a new position in swimming; the vertical plane of its body slopes more and more from right to left as the eyes shift round so that in swimming the plane passing through both eyes is always horizontal. At the completion of metamorphosis the whole symmetry of the head has been profoundly disturbed, though that of the body remains as before, except that the opening of the ureter has shifted from the median ventral line to the right side. The horizontal swimming plane of the whole body has been rotated through a right-angle and the fish rests and swims on its morphological left side. The pigmentation now gradually disappears from the lower side.

The larvæ now feed almost entirely on Copepoda, with which their stomachs are usually crowded; larval Molluscs and larval Crustacea are also eaten. After the metamorphosis the food is changed, and the small fishes, $1\frac{1}{2}$ to 4 inches, feed largely on various Annelids such as *Nereis* and *Pectinaria*, on small Crustacea such as *Mysis*, and various Amphipods and small *Crangons*. Later on the fish adopts its definitive food, which is mostly Mollusca, the favourite forms being *Cardium*, *Tellina*, *Mactra*, *Scrobicularia*, *Donax* and *Mytilus*. The character of the food changes little during the rest of its life.

Rate of growth.—A variable time is taken up by the changes above described, which were observed in larvæ kept in aquaria. For instance, although the Plaice in the Danish seas may spawn as early as November, no larvæ of 12mm. length are taken there till May. They therefore require six months to pass through the metamorphosis. Two methods have been adopted for estimating the rate of growth subsequent to the larval period. The most obvious one is to keep a large number of young fish in captivity in aquaria, and to observe for as long a period as may be possible the changes in body length. This method is evidently open to the grave objection that the fishes live under artificial conditions, and these may influence their rate of growth. The other method is to fish often in waters which contain great numbers of Plaice of different sizes, and to deduce the growth rate from the grouping of the individuals of different lengths. When this is done and the results of measurement of all the fishes captured are tabulated, it is seen that those captured at any one time may be arranged into several groups in each of which the greater number are grouped round certain typical lengths, and if such experiments are kept up for some time an approximation to the rate of growth may be arrived at.

Such experiments have been carried out both in this country and in Denmark. Dannevig* found that on the East Coast of Scotland the Plaice grows about three inches in the year. The average size at the end of the first year is 77·2mm., at the end of the second 156·7, and at the end of the third year 233·1mm. The experiments also show that it is during the warmer part of the year—May to October—that growth takes place. During the winter it is practically arrested.

The period following the metamorphosis of the fish till the third or fourth year may be called the immature stage. It is characterised by the undeveloped condition of the reproductive organs, and by the gradual outward migration of the fish towards deep water. The ovaries remain small and contain only small transparent ova. At a variable size and age sexual maturity is indicated by the gradual enlargement of the ovaries and testes and by the occurrence in the former of eggs containing yolk. After the first sexual maturity occurs the ovaries never revert to their condition prior to maturity, and it is always possible to distinguish a fish which has previously spawned.

Sexual maturity.—The size at which first-sexual maturity occurs is variable within somewhat wide limits in the seas round the British Isles, and as a knowledge of this size is of considerable importance as a basis for fisheries legislation, great pains have been taken to estimate it for various localities. It may vary within very wide limits—thus in Danish waters spawning female Plaice of 7″ in length have been exceptionally taken,

* On the rate of growth of the Plaice. 17th An. Rep. Scottish Fish. Bd., p. 232, 1898.

while in the North Sea immature female fish of **17″** have been observed. In the male sexual maturity is first attained at a lesser size than in the female; in Danish waters ripe males 7 inches long, and in the North Sea 9 inches occur. The most convenient size* for purposes of legislation is that lowest size at which as many Plaice are mature as immature. This is the average size of first maturity, and it is of course lower in the male than in the female. In the northern part of the North Sea the female reaches maturity for the first time on the average at $15\frac{1}{2}$ inches, and the male at $12\frac{1}{2}$ inches. In the southern part of the North Sea these sizes are $13\frac{1}{2}$ inches for the female and $10\frac{1}{2}$ for the male. In the English Channel Cunningham found that nearly all Plaice were mature at 15 inches, and in the North Sea Holt found that at 17 inches nearly all the fish observed were mature. In the Danish seas much lower sizes than these have been estimated by Petersen.† Thus the average sizes at which Plaice first become mature are 10 inches in the Baltic, 11 inches in the Lesser Belt and 12-13 inches in the Kattegat. In the Irish Sea the smallest mature female observed was 13 inches in length, and the largest immature female 19 inches. The smallest mature male obtained was $10\frac{1}{2}$ inches long, and the largest immature male 13 inches. Sufficient fish have not been examined in this district to enable satisfactory average sizes for the first maturity to be established.

It is probable that the Plaice spawns every year after sexual maturity is attained, but there is no certain evidence on this point, nor is it known what are the yearly increments of growth after the third year of life. In the

* See Kyle, 18th An. Rep. Scottish Fish. Bd., Pt. iii., p. 189, 1900, for a discussion of this subject.

† 4th Report of the Danish Biol. Station, 1894, p. 3.

P

largest fish obtained the reproductive organs are still functional. The maximum size for the North Sea seems to be
28 inches, for the Danish seas it is smaller, about $22\frac{1}{2}$
inches. When the Iceland Plaice fisheries were first
exploited, very large specimens—over 30 inches in some
cases—were obtained.

Migration and distribution.—Wherever the distribu
tion in space of the Plaice has been attentively studied it
has been found that a very well marked segregation in
respect of the size of the fish exists. There is a close
correspondence between the size of the fish and the depth
of the water at the bottom of which it is found. The size
increases with the depth. The fish does not exist outside
the hundred fathom line, and indeed practically all the
fishing is carried on in seas varying from 20 to 50 fathoms
in depth.

Since only the larger individuals are functionally
reproductive, it follows that the Plaice spawns only in
deep water, and it is the case that the spawning grounds
are always situated at some distance from the shore.
These spawning grounds are usually very definitely
located in any district. Thus on the East coast of Scotland such areas occur in the North about 16 miles seaward
from Moray Firth, and further South off St. Andrews Bay
and the Firth of Forth. In the Danish seas the fish only
spawns in the more open waters, such as in the eastern
parts of the Kattegat, in the Belts and in the Baltic. In
the Irish Sea the positions of some such spawning grounds
have also been determined, and one notable area lies East
from the South end of the Isle of Man in a depression
having an average depth of about 23 fathoms and surrounded by water of 16-20 fathoms in depth. The bottom
consists of soft bluish-black mud with an abundant fauna.
Scrobicularia and *Turritella* are very abundant, and the

Pennatulid *Virgularia* is also common and forms a constant food of various Gadidæ. Many edible fishes spawn here, Pleuronectids like the Plaice, Flounder and Dab, and Gadoids such as the Cod, Haddock and Whiting. During the spawning season ripe fish may always be taken by trawling on the ground, and pelagic eggs in various stages of development are found in plankton gatherings made at the surface. Another spawning ground exists further North, due East of the North end of the Isle of Man. On the West side of the Isle of Man are grounds due West of Dalby where other Pleuronectids, and probably the Plaice also, spawn, and it is very probable, though systematic surveys have not yet been made, that similar spawning areas lie further South out from the Lancashire and Welsh coasts.

As we have already stated the Plaice emits its spawn at the bottom of the sea and the eggs rise towards the surface. Here their inshore migration begins. The developing embryo while still within the egg capsule has of course no power of movement of its own, and even after it has hatched and is a pelagic organism its powers of migration are extremely limited, so that its present movements are determined entirely by the combined operation of physical agents, waves, prevailing winds, tidal drift and currents, and from a consideration of the working of these factors the general course of the eggs and embryos from the spawning grounds can be determined.

The general drift of small objects floating at or near the surface of the sea has been studied in the Irish Sea by observing the movements of weighted bottles designed to drift partially or wholly submerged. Two such series of experiments have been made,* and their results shew that the combined operation of the various agents indicated

* Lancashire Sea Fish. Laby. Reports, 1895, p. 12, and 1898, p. 30.

above is to carry small floating objects in an easterly and northerly direction. Thus eggs and embryos floating at the surface on the spawning ground to the South-east of the Isle of Man will travel easterly with a slight drift to the North to Morecambe Bay, the estuary of the Duddon and to the Cumberland coast. It is also evident that the young fish inhabiting the shallow waters further South, that is in the Ribble and in Liverpool Bay and the Cheshire coasts, must have originated elsewhere than in the spawning grounds referred to. Mr. R. L. Ascroft informs us that from his experience of the drift of floating wreckage it is probable that fish spawning far South in Cardigan Bay and off the Welsh coast generally produce the young fish in the southern and central portions of the Lancashire coast. Up to the present time the spawning areas off the Welsh coast have not been investigated.

This first pelagic stage of the young Plaice is the period during which both embryonic development and larval metamorphosis take place. Up to the time when the rotation of the eyes is fairly in progress the young fish swims freely in the upper layers of the sea, and during that change it gradually sinks until, when metamorphosis is completed, it finally settles on the bottom. It is generally agreed that the young Plaice during this change cannot, or at least does not, inhabit the sea at any great depth. It is probable therefore that should the larvæ begin to sink while still in deep water their destruction follows, and it seems essential that they should have nearly completed their larval changes not earlier than the time when they arrive at the shallow coastal waters. About 30 days after hatching the larvæ seek the bottom and this gives a period of over 40 days for them to complete their inshore drift from the spawning grounds.

During this drift inshore the embryos and larvæ are

naturally exposed to many dangers. They are probably eaten by many pelagic animals, though there is not much published evidence on this point. In Loch Eyne, however, the floating fish eggs are closely associated with great numbers of Copepoda, and this results in the eggs being eaten by the herring in the search of the latter after Copepoda and other pelagic Crustacea, and pelagic fish eggs have been found in the stomach of that fish. It is possible, however, that physical events are at least as fruitful causes of the destruction of floating eggs and larvæ as predaceous pelagic animals. The change from the pelagic to the demersal mode of living, for instance, happening when the larva is still in deep water. Remarkable conditions are present in the Baltic. Petersen* has shown that young Plaice (up to 2-3 inches in length) are entirely absent in that sea, though spawning fish are abundant, and, as Hensen has shown, fertilized eggs are there in enormous abundance. The low specific gravity of the water affords the explanation. At a specific gravity of 1·0140 (at 9°C.) a great number of Plaice eggs sink to the bottom, and at a specific gravity of 1·0120 (at 10°C.) all sink. The lowest specific gravity at which the eggs can drift about without any sinking is 1·0152 (at 9·8°), and if in their migration they enter water of less than this density their destruction follows. Now it happens according to Hensen that about once every month there occurs such a low specific gravity of the Baltic water that all the Plaice eggs sink. There is of course a possibility that the eggs may go on developing at the bottom, but this is unlikely. It is possible too that a low salinity of the water may prejudicially affect the processes of development, but this subject has not been adequately investigated. It is possible also that Plaice eggs entering the estuaries on the Lancashire coast may

* Rep. Danish Biological Station, IV., 1894, p. 27.

be destroyed in this way, but the variation in the salinity of those waters has not been thoroughly investigated.

Wherever spawning grounds may be located in the Irish Sea, it is hardly likely that a longer period than 40-45 days can elapse before the larva finds itself in suitable conditions on shallow sandy shores.

The young Plaice now enters on its life in the "nursery" ground. Practically the whole of the Lancashire and Cheshire coast consists of ground of this character—flat, sandy shores with shallow water overlying. There are, however, several localities to which in particular the term is applied; there is the whole of the shallow water round the mouth of the Mersey, the whole estuary of the Ribble; the ground out from Blackpool known as the Blackpool closed ground, the whole of Morecambe Bay and the estuary of the Duddon at the boundary of Lancashire and Cumberland. On all these grounds vast quantities of young and immature Plaice up to fish of 2-3 years old are found. Very young Plaice and other Pleuronectids appear on these shores during June every year. They have as a rule completed their metamorphosis, though a few may generally be got with the eyes in the course of rotation. They are generally about $\frac{1}{2}$ inch and less in length, but have assumed the perfect form of the adult (fig. 37). They may be found in great numbers in the shallow sandy pools left by the receding tide, where very many are stranded and perish. By the autumn these little fish have grown to about 2-3 inches in length. They are then present in great numbers on the banks or shallower waters of the nursery ground. In winter they disappear to a great extent from these banks, or at least are not taken in the shrimp trawl, and it is probable that they either bury themselves in the sand when the colder weather approaches (this is a common thing with much older Plaice), or they migrate to the deeper portions of the nursery ground such as the channels. The latter seems to be the case with the Mersey nursery. Great quantities of young Plaice are to be found there on the sandbanks, where the water is shallowest, during July, August and September, maximum quantities being taken in the latter month. During the winter months, however, comparatively few are found there, but they are abundant in the channels where the water is deeper. This local migration then goes on independently of the larger movement. The Plaice move from the shallow water to the deeper as the colder months approach and from the deeper water in the channels to the shallow banks as the temperature rises.

The young Plaice on these nursery grounds form part of an exceedingly abundant vertebrate and invertebrate fauna. They are associated with other Pleuronectid and Gadoid fishes—the dab, flounder, sole and solenette (*Solea lutea*), with occasionally young brill and turbot and the whiting, haddock and cod. Young sprats (*Clupea sprattus*), sand-eels (*Ammodytes*), *Cottus*, sting-fish (*Trachinus*) and *Centronotus* are also found; all these with the exceptions of the sting-fish, sprat and solenette are young and immature fish. Of the Gadoid fish the whiting are very abundant. These are young fish, in their first year probably, and generally not exceeding five inches in length. The invertebrates are usually crabs (*Portunus*) and star-fish (*Asterias*), and great numbers of shrimps (*Crangon*). The crabs alone often form nearly half the total bulk of the catch.

These young fish are continually being captured in the shrimp trawl, which having a square mesh of ½ inch side, retains them. We may quote one single catch to give an idea of the bottom fauna of the Mersey nursery ground associated with the Plaice during August. The catch was taken by a shrimp trawl the length of the mouth of which was 25 feet and with a mesh of $\frac{1}{2}$ inch side. It was dragged for one hour over two miles of ground in six fathoms of water. The bottom was a mixture of sand and mud; the contents of the net were—

Soles, 257; 11 were over 8 inches long, $\frac{1}{3}$ of the catch were 2″-8″, the remainder 2″.

Plaice, 265; 6 were over 8″, remainder 2″-8″

Dabs, 896; 2 were over 8″, remainder 2″-4″.

Skates, 18; 7″ broad across pectoral fins.

Whiting, 285; 5″ long.

Shrimps, 20 quarts (a quart contains from 200 to 400 animals).

In addition to these about 200 Solenettes were caught, many other fishes (*Trachinus*, *Ammodytes*, *Clupea sprattus*) and a great number of Crabs.

On the Blackpool Closed Grounds in the central part of the Lancashire shores even larger catches of young Plaice have been made. Thus in 4 drags with a shrimp trawl on—

September 25, 1893, for a drag of $2\frac{1}{4}$ miles, 3,302 plaice were caught.
December 28, 1893,   „   „ $1\frac{1}{8}$ „   4,520   „   „
January   2, 1894,   „   „ $\frac{1}{3}$ „   2,511   „
February 14, 1894,   „   „ $\frac{1}{2}$ „   2,151   „   „

Over 10,000 young Plaice, of 2″-4″ mostly, being caught in these four drags alone.

In the various nurseries the fauna associated with the Plaice varies somewhat, but in all shrimps are always found where young Plaice and other Pleuronectids occur.

With increasing size the immature Plaice move out from the nurseries into deeper water. This off-shore migration has been studied in an ingenious way by

Fulton* on the East coast of Scotland. Spawning takes place on the off-shore grounds out from the mouths of the Forth and St. Andrews Bay, and the eggs are borne inwards and supply the nurseries in those waters. Numbers of Plaice on these inshore grounds were captured and marked by the attachment of a numbered label, and then liberated. After variable periods these fishes were re-captured, generally by the fishermen, and then returned to the Fishery Board officers. In this way the course after liberation was determined, and it was found that a slow migration from the South and round the North coasts of the Firth of Forth and then round Fife Ness into St. Andrews Bay took place, the fishes then moving outwards from these shallow waters to the spawning grounds. It is certain that some such movement takes place on the Lancashire coasts, though Fulton's experiments have not been repeated there. The distribution of sizes is, however, sufficient evidence, backed with what we know of the actual movements in other places. We have seen that on the nursery grounds great numbers of young Plaice of about $\frac{1}{2}$ inch long are found during the early summer, and owing to the great quantity of these small fishes the average size at that time must be very low. Towards September, however, these little fishes have entirely dis-appeared from the sandy pools, and the numbers of Plaice on the grounds slightly off-shore (up to 6 fathoms in depth) have greatly increased. These fishes, which are now from 2 to 4 inches in length, are the same individuals which crowded the sandy pools between tide marks some months earlier, and the first part of their off-shore migra-tion has begun.

Further out to sea within territorial waters generally the average size of the Plaice caught in the trawl nets is

* 11th An. Report Scottish Fish. Bd., p. 176, 1892.

much greater. Probably about 9 inches will represent the average size of the fish caught in Morecambe Bav and generally within the 3 miles limit. These fishes are now marketable, and in suitable places all along the coast great quantities at this size are caught in the stake nets. Larger individuals are of course often got, but on these in-shore fishing grounds mature Plaice are never or only very exceptionally captured. It is these immature fishes which in their gradual migration outwards form the mature Plaice population of the more open sea. At or near the spawning season they congregate on the spawning grounds.

Briefly summarized then the migratory movements of the Plaice are (1) the passive drift in-shore of the develop ing eggs and metamorphosing larvæ terminating as the larva acquires the adult form and settles to the bottom, and (2) the slow outward movement of the young fish, deeper water being continually sought as it increases in size. This movement ends as the fish becomes mature. Thereafter its movements are probably very limited. During the spawning season it probably does not travel at all. With the extrusion of spawn another generation begins the migratory cycle.[*]

---

[*] We have described only the larger migrations which are part of the life movements of the Plaice. Smaller and local migratory movements are of course continually going on. Mr. R. L. Ascroft informs us of a curious instance which illustrates the connection between these smaller movements and physical events. A severe storm about 1885 was followed by a very marked increase in the numbers of Plaice in one of the channels of the Ribble estuary—the Bog Hole. For about four days great quantities were caught, one of the sailing trawlers getting as much as 180 score of fish (value £30) in a day. The cause of this remarkable abundance was that the storm and rough water washed off the " Sand pipes " (*Pectinaria belgica*), which existed in great abundance on the neighbouring banks, into the channel, and the Plaice followed the food.

B.—The Plaice Fishery.

We are able to present here only a very brief sketch of the economy of the Plaice, and the reader who is sufficiently interested in this direction is referred to the very extensive literature which has now been published in Britain, Germany and Denmark. We propose to deal only with the methods in use for the capture of the Plaice, and with the causes of the apparent decline in value of the fishery, and the regulations which have been suggested for the future welfare of the industry.

Round the British coasts the Plaice is fished for in three ways. It is caught by spearing, by means of stake nets and by trawling. Of these methods the latter is incomparably the most important. Comparatively few fish are caught by spearing, and this method is only pursued in shallow waters and then to a very limited extent. Stake-net fishing is much more important, and is carried on at many parts of the coast on the foreshore between tidemarks where conditions are suitable. The net is about a yard in width and is of variable length. In Lancashire waters it may not exceed 300 yards in length, and it has a square mesh of either 6 or 7 inches in periphery. It is stretched on a row of stakes driven into the sand in a straight line, at right-angles to the direction of the tidal flow. It may be provided with pockets. Fish swimming with the tide are caught in the meshes and are removed when the net is "fished" at low water. At sea the Plaice is fished by means of the trawl. The trawl net is a conical bag of variable length. Its mouth is held open by means of a wooden beam on which the net is stretched, so that the open mouth is rectangular in shape. At either end of the beam are fastened iron frames, the "irons," the lower parts of which rest on the ground. The lower margin of the mouth of the net, that which rests on the ground, is laced on to a stout rope—the "foot rope," which is much longer than the beam and accordingly drags behind the latter in a wide bight. From either iron a strong rope proceeds—the "bridles," the free ends of which are fastened together by the "shackle." To the shackle the trawl warp, either of rope or of steel strands, is fastened. When fishing the whole contrivance is dragged on the sea bottom, and after a variable time the net is hauled on board the vessel, and the apex of the cone, the "cod end," is untied, and the contents are allowed to drop out on deck.

The dimensions of the trawl vary. In Lancashire territorial waters (within 3 miles from low-water mark), when the length of the beam does not exceed 18 feet, there must not be less than 50 rectangular meshes in the circumference of the net, when it exceeds 18 and is less than 25 feet there must not be less than 60 meshes, and when the beam exceeds 25 feet the net must contain not less than 80 meshes. The mesh of the trawl net must measure (with a certain exception) 7 inches at its periphery. Within the territorial waters only sailing vessels may trawl. Outside on the high seas there are of course no regulations, and the trawl may have any form and dimensions desired. When employed by steam fishing vessels the trawl beam may be as long as 50 feet, but the beam trawl seldom exceeds those limits.

Since about 1896 the beam trawl has been largely superseded in deep sea trawling vessels by the "otter" trawl. In this apparatus the general form of the beam trawl is retained, but the beam is discarded, and its place is taken by a strong rope, the "head line," to which the upper margin of the mouth of the net is laced. The foot rope is the same as in the beam trawl but the head line may be 120 feet long. The mouth is kept open by being attached to " otter boards," which are very large and heavy wooden boards shod with iron, one edge of each of which rests on the ground. They take the place of the irons in the beam trawl. They are set at an angle to the direction in which the net is dragged, and by being pressed outwards when pulled keep the net open. The net is hauled by two warps, one of which is attached to each otter board and passes in over one of the quarters of the vessel. The otter trawl is stated to have an efficiency 37-50 per cent. over that of the beam trawl.

These are the two principal methods of fishing in this country. In Denmark the " seine " net takes the place of the trawl. The Danish Plaice seine is a bag of netting about 20 feet long, the mouth of which is produced out into two wings of about 180 feet in length. The depth of the mouth and wings is about 7 feet. There are 5 or 6 meshes to the foot in the netting. The apparatus is used from an anchored vessel by being " shot " in a wide curve at some distance from the vessel. It is then hauled by two warps, one attached to each wing. The net drags on the bottom in the same manner as the trawl. In Denmark the fish are landed alive, a custom which is quite exceptional in British fishing, the fish being brought to the port of landing preserved in ice.

What the real value of the Plaice fishery in British waters may be is difficult to determine accurately. The official returns made by the Board of Trade collectors of statistics show that for the year 1898, 35,788 tons of Plaice having an initial value of £873,680 were landed at British ports. Of this total quantity 31,544 tons were landed at East coast ports, 2,355 tons on the South coast, and 1,882 tons on the West coast. It will be seen how important the East coast Plaice fisheries are, those grounds yielding over 88 per cent. of the whole, while the South coast yields 6 per cent. and the Irish Sea only 5 per cent. It is generally believed, however, that these official returns of the Board of Trade are imperfectly collected, and the above figures are to be regarded as minimum values. Accurate returns of the numbers of fish landed at Grimsby in one year alone are furnished by Holt.* For the year ending March, 1894, over $17\frac{1}{4}$ millions of fish were landed at that port. Most of these fish were caught in the North Sea, a small proportion only in Iceland waters.

We have seen that the Plaice, like other Pleuronectid fishes, is a permanent bottom living fish, has a comparatively limited distribution, and a range of migration which is relatively small. On account of these habits it is peculiarly liable to capture by present means of fishing, and at no stage in its life history from the metamorphosis onwards does it go outside the range of the fisherman's operations. While yet on the nurseries it is caught by the shrimp trawler, on the inshore grounds it has become marketable and is caught in the stake and trawl nets, and in the open seas, when mature, it becomes the prey of the deep sea trawler. In many of these respects it contrasts with the commoner round fishes, such as the herring, cod or mackerel. While it is generally agreed that the supply of the last named fishes is practically inexhaustible, it seems no less clear that the abundance of Plaice and other flat fishes may be very sensibly influenced by the operations of modern fishing, and in fact many arguments point to the conclusion that the flat fisheries on British coasts are declining. The present system of collecting statistics is, however, so incomplete that to make out an absolutely convincing proof of this is difficult, and it is only fair to

* Holt—An Examination of the present state of the Grimsby Trawl Fishery. Jour. Mar. Biol. Assoc., vol. iii. (N.S.), 1893-5, p. 339.

state that contrary opinions have been expressed. It is at first sight a paradox that year by year the total catch of fishes in British waters should increase, while the fishing grounds may be really deteriorating. Along with this increase in total quantity of fish caught, however, has gone on a marked increase in the catching powers of the fishing fleets and an extension of the area fished over. The introduction of steam into fishing vessels about 1850, and the use of ice for preserving the catches, made possible the use of larger and more efficient apparatus (the otter trawl latterly), and enabled the vessels to make longer voyages. With this change the small sailing boats began to decrease in numbers, and fishing instead of being carried on by vessels independently owned by masters or crews, became a great capitalised industry.

Therefore although the annual catch has gradually increased, the average catch per vessel is now beginning to decrease, and the density of the fish population in the seas round the British Isles is less than it formerly was. The catches of 4 Grimsby sailing trawlers for every year since 1875 have been published by Garstang,[*] and all of these shew a marked decrease with hardly any fluctuations.

In Danish seas the same decline has been noticed. Petersen[†] has shewn that there has been a steady decrease in the size of the Plaice landed for many years, and it is obvious that the reduction in size of the fish landed is indicative of the reduction in number of those present on the fishing ground.[‡]

[*] See Garstang Mar. Biol. Assoc. Journ. for these figures and the elaboration of the above argument.

[†] 4th Rep. Danish Biological Station, 1893.

[‡] From the relation between the average size at first maturity and the average size of all mature plaice captured, the change in the fish population of an area may be deduced. See Kyle—18th An. Rep. Scottish Fish. Bd., 1900, p. 200.

It is worthy of note, however, that the decline in the annual quantity of Plaice taken from the North Sea may be regarded as the reduction of an "accumulated stock"* and not entirely as the diminution of the number of fish annually coming to marketable size. At one time the North Sea was practically a virgin ground, and the large and numerous fish taken from it were a stock which had accumulated for many years and which were rapidly fished out.

Obviously the estimation of the fluctuations of the fish population of a great fishing ground is a task of the utmost difficulty. We may mention the quantitative method of plankton investigation as one of the most promising means of dealing with this problem. This method has been greatly developed during recent years by Hensen and the German Fisheries investigators, and it is now possible to make a rough estimation of the number of pelagic fish eggs of any species such as the Plaice, in an extended fishing ground. Only a rough approximation of the eggs present can, of course, be made since many difficulties and sources of error are obviously to be considered. But the Hensen method is as yet the only serious attempt at a "census of the sea" which has been attempted, and it is possible that its refinement and thorough application may go a long way towards solving the question of fisheries impoverishment. We have stated that Hensen estimates the number of Plaice eggs floating in the North Sea during 1895 as about 31 billions. Now it is known that a mature female Plaice produces annually about 300,000 eggs and it was then easy to calculate that about 103 millions of mature female Plaice were present in the North Sea in that year. From the known ratio of the

* See Hjort and Dahl, Rep. on Norwegian Fish. and Mar. Investigations, vol. i., no. 1, p. 151, 1900.

sexes the total number of mature fish can be further calculated. Such determinations made from year to year
would obviously give the fluctuation of the adult Plaice
population.

Of all the Pleuronectid food fishes the Plaice is the
most important by reason of its relatively great abundance
and on that account when the probability of the decline of
the fishery became evident it was the object of much solici-
tude on the part of the fisheries authorities, and many
remedial measures have been proposed. In this country
the continual destruction of immature fish has been the
most obvious danger to the fishery, and the remedies have
all been directed to the minimising of this evil. The idea
underlying the remedies suggested has been to allow as
many fishes as possible the chance of spawning at least
once in their lifetime. This destruction of immature fish
takes place in every form of fishery. Practically all the
fish captured in inshore fishing, whether by trawl or stake
net, are immature. Even in the deep sea Grimsby fishing
more than half the fish landed are immature. In 1894[*]
of the Plaice landed at that port 7 millions were mature
and 9 millions immature. Of the same quantity $9\frac{1}{2}$
millions were over 13in. in length and $6\frac{1}{2}$ millions under.
At first sight therefore it might appear that the imposition
of a size limit below which it would be illegal to land the
fish would be an effectual remedy. If based on the size at
which sexual maturity first occurs this limit would vary
for different localities. It is obvious, however, that such
a regulation is impracticable, since the Plaice becomes a
marketable fish long before it becomes sexually mature.
The imposition of such a limit would close the inshore
grounds against plaice fishing altogether.

Other size limits have, however, been proposed, and

[*] Holt—*loc. cit.*, p. 410.

Q

the adoption of some one will probably come in the future. In 1892 a conference held under the auspices of the National Sea Fisheries Protection Association proposed 10in. as the legal minimum size of marketable Plaice. Following this, in 1893 a Select Committee of the House of Commons recommended 8in. as the limit, being influenced by the existing legal sizes in other countries. In France the legal size of Plaice is 5½in., in Belgium nearly the same, and in Denmark it is 8in.

It is to be noted that the arguments as to the decline of the Plaice fishery are generally founded on the total *weight* of fish landed, not on the number of individuals, and this has suggested the ingenious " growth-theory " of Petersen* for the betterment of the fishery. In Danish waters the weight of a 10in. Plaice is less than ½lb., and that of a 14in. fish is more than twice as much. Now if in the time required for a Plaice to grow from 10 to 14 inches the total mortality on the fishery ground is such as to reduce the number of individuals by one-half, the total weight of fish will remain much the same as before. But it is not likely that the mortality will be so great, since the Plaice is singularly free from disease, and at the sizes mentioned its enemies (predatory fishes) are few. It follows then that by deferring their capture until they have grown to 14in. in length a much greater weight of fish will be brought to the market. The utility of a size limit (the most profitable one to be determined by experience) is therefore apparent. Though fewer fish are caught the fishing will become more profitable. The practicability of such a measure is greater in Denmark than in Britain. There fish are brought to the market alive and instruments of capture are designed to that end. Small fish taken can therefore be returned to the sea alive. Here,

* Petersen—*loc. cit.*, p. 48.

with large trawls, heavy catches and long drags, most fish are dead when the net is hauled. The success of such a measure would therefore depend to some extent on the possibility of devising a mesh of such form and dimensions as would capture only moderately large Plaice without prejudice to the capture of other (round) fish trawled for.

As far as inshore fishing is concerned the "vitality" experiments made by Mr. Dawson[†] for the Lancashire Sea Fisheries Committee have shewn that with moderately short drags (one or two hours) both with fish and shrimp trawls a great proportion of Plaice (81 per cent.) recover when taken from the contents of the net and placed in running sea water. The relative catching powers of nets with different sizes of meshes have also been ingeniously illustrated by the same writer. An ordinary fish trawl of 25 feet beam and with a square mesh of 7in. periphery was used. Round the catching portion of this net a similar net but with a mesh of $4\frac{1}{2}$in. periphery was laced. The combined net was dragged in the ordinary manner. Fishes which passed through the inner wide meshed net were retained in the outer one. In one such trial the inner 7in. net captured 41 Plaice of about 9 inches in length, while the outer $4\frac{1}{2}$in. net caught 349 Plaice of about $5\frac{1}{2}$ inches in length which had passed through the inner net. In another trial of this combination net the 7in. net caught 142 Plaice $7\frac{1}{2}$-$9\frac{1}{2}$ inches long, and the outer $4\frac{1}{2}$in. net 390 fishes of $4\frac{1}{2}$ inches long. Corresponding results were obtained with the other fishes captured.[‡]

Petersen's "growth-theory," it will be seen, proposes to remedy the exhaustion of the Plaice fishery by raising

[†] See Lancashire Sea Fish. Laby. Rep. for 1893, p. 23.

[‡] See Holt—Journ. Mar. Biol. Ass., vol. iii., pp. 437-441, 1895, for a discussion of the probable results of regulation of the trawl mesh.

the size and with this the weight of fish landed. The author contends that there are already sufficient eggs and fry in the seas, and that it is not useful to attempt to add more. In this country the view has generally been that the decline in the fishery can be remedied by increasing the quantity of eggs or fry in the nurseries, either by increasing the number of spawning fish by the imposition of a minimum size limit or by artificial incubation, or by protecting the very young fishes on the nursery grounds. In the Irish Sea the nursery is also a shrimping ground, and the only means of attaining the latter object is by interference with the shrimp trawl fishery, for the results of experimental shrimp trawling which we have quoted above sufficiently demonstrate that enormous destruction of Plaice fry necessarily accompanies shrimp trawling in most places where it is carried on. Various attempts have been made by the Lancashire Sea Fisheries Committee to obtain powers to legislate in this direction, and the area known as Blackpool Closed Ground is as yet the only fishing ground where shrimp trawling is forbidden in the interests of the young Plaice and soles which are reared there. The amount of destruction of young Plaice which was due to shrimp trawling on that area will be seen from a consideration of the figures we have quoted above.

The shrimping ground at the mouth of the river Mersey is an area on which shrimp trawling is extensively practised,[*] and where as we have seen great numbers of young Plaice unfortunately congregate. Experimental hauls with a shrimp trawl have been made for many years by the officers of the committee on a portion of this area, which the Committee recently unsuccessfully attempted to close against trawling during July, August and

* See Lanc. Sea Fish. Lab. Report for 1900, p. 39, 1901, for a short account of this fishing ground.

September of every year. As the result of 7 years' experimental fishing it was found that 567 young Plaice were taken in an average haul with a shrimp trawl on this ground in those months. Now during those 3 months 15 shrimp trawling boats on the average are fishing there every day, each boat making 3 hauls per day on five days per week, 2,925 hauls in all. Supposing each boat to have made the average catch of 567 young Plaice, over one and a half millions of young fishes would have been caught on this area alone during the three months mentioned per year. It must be remembered that in the fishing as ordinarily practised the great majority of these are really destroyed.

The enormous destruction of young fishes due to this method of fishing will be realised when we state that there are altogether about 100 boats employed in shrimp trawling in the Mersey estuary, and the grounds seaward from the river, and that the above calculation applies to only 15 of these frequenting a particular area for 3 months. The Mersey is not the only district in which shrimp trawling is practised. Just as active fishing is carried on in the Ribble estuary and in Morecambe Bay and in many parts of the coast, "cart-shanking"—an equally destructive form of fishing, is practised. We believe we are under the mark in stating that the yearly destruction of young Plaice on the Lancashire and Cheshire coasts before the closure of the Blackpool ground due to shrimp trawling must be measured by the hundred million.

It is a regrettable circumstance that regulation with a view to the protection of the young Plaice on the nursery grounds must to some extent interfere with the proseoution of shrimp trawling, but it seems probable that the loss incurred by the latter fishery would be more than compensated by an increased number of Plaice on the fishing grounds, and even if these were caught while still immature fishes frequenting territorial waters there would still be a net gain. It is a question worthy of consideration therefore whether, in the interests of the Plaice fishery, some restrictive measures applied to methods of shrimp fishing would not be of benefit to the whole fishing industry.

We are referring here to the method of fishing for shrimps with the shrimp trawl, which is a net of the pattern of the larger fish trawl described above but with a mesh of 2 inches in periphery. Other forms of shrimp nets are used, particularly the " shank " net, which is a net like a trawl but having the beam, foot rope and irons replaced by a rectangular wooden frame. A further improvement consists in attaching the lower edge of the net, not to the lower bar which drags on the ground, but to another bar, parallel to this but placed two or three inches above it. When disturbed the shrimp jumps, and, clearing the bar, enters the net, while the fish when disturbed swims off close to the ground and may escape through the space between the two bars. Experiments have shewn that these two nets capture relatively less fishes and more shrimps than the ordinary shrimp trawl.

We have yet to refer to the artificial incubation of Plaice eggs as a means of recruiting a fishing area in process of exhaustion by overfishing. So far this has only been extensively practised in Scotland. Mature male and female Plaice are captured some time before spawning begins and are penned in a " spawning pond." Spawning and fertilization take place in the pond as in the sea and the fertilized eggs are collected daily and are put into the " hatching boxes." Through these a constant stream of sea water passes and they are kept in continual motion to avoid the clustering of the eggs and the risk of insufficient aeration. The eggs develop in these boxes and the larvæ after being retained for some time are taken out to sea and "planted" in a suitable locality. It will be obvious from what we have stated above with regard to migrations that the larvæ must be set free in such a place that the prevailing winds and tidal drift will carry them to a nursery (which must be present in the area dealt with) in which the conditions for further development are suitable and in such a time that the assumption of the demersal habit will coincide with the arrival of the larvæ on the nursery grounds. The ultimate aim of the hatching operations is to rear the larvæ through the period of their metamorphosis under artificial conditions. It has, however, been found extremely difficult to rear even a small proportion through the period referred to since great numbers die during the period immediately following the absorption of the yolk sac. In practice, therefore, the larvæ are set free before this mortality has seriously commenced. The Scottish Fishery Board, in addition to their utilitarian object of adding to the number of young fish in the sea, have since 1897 been devoting attention to the interesting experiment of placing some millions of fry yearly in one limited area where they have reason to think they may be able to test the result by periodic observations on the young fish fauna. In the period between 1894 and 1899 inclusive the Board set free 136 millions of artificially hatched larvæ, and since 1897 these have been planted in Loch Fyne, the area chosen for experiment.

The argument for the utility of sea fish hatching rests to some extent on the hypothesis that the period during which the fish are being dealt with in the hatchery is that during which the mortality in nature is greatest. Since, in the hatchery, the eggs and larvæ are safe from enemies or prejudicial changes in their physical surroundings, this mortality is avoided. It is most probable that during their pelagic life the eggs and larvæ are destroyed by being eaten by other animals or by physical changes, but unfortunately we have as yet no idea of the proportion of eggs or larvæ so destroyed. Experience in the hatchery shews that it is during the period between the absorption of the yolk sac and the beginning of the metamorphosis that the mortality is greatest. This critical period is characterised by the change in the means of nutrition of the larva, which having used up the food yolk, begins to feed on planktonic organisms. If this mortality exists in nature, and if it should become possible to avoid it in the hatchery, then the gain would be very considerable, but until it shall become possible to rear the greater portion of the larvæ hatched through their metamorphosis all that is gained in the hatchery is the immunity of the eggs and larvæ and of the fishes in the spawning pond. From this latter point of view—the immunity of the fish yielding the eggs—the hatchery is to be regarded as a reserve of spawners, and its function becomes the more valuable the greater the reduction of the fish population in the area dealt with. It is to be regarded as effecting the same end as would be brought about by protection of mature fish on the spawning grounds—protection which at present seems impracticable, or protection of the fry caught in the course of other fishing—protection also apparently impracticable. We refer here to the treatment of restricted areas. Probably the effective treatment of such an area as the whole North Sea by artificial hatching is at present impracticable. The deficit which the hatchery would have to make good is the assumed reduction of the mature Plaice population, and this most probably takes place on a scale which it would be difficult to approach by artificial operations according to our present ideas and methods.

It has been suggested that much the same results would be attained by the removal and fertilization of the ripe eggs from Plaice caught in the course of commercial fishing and the immediate return of these to the sea without having undergone treatment in the hatchery. We do not know whether this is practicable, and at any rate, the number of ripe eggs in the ovary of a Plaice at any one time is only a small proportion of the total number present. This method of obtaining eggs for the hatchery has, however, been adopted at times. The trawling vessels have been boarded on the fishing grounds and ripe fish have been "stripped," the eggs fertilized and taken to the hatchery.

As we have seen the most striking fact in the life history of the Plaice is its limited range of migration, and this has suggested the possibility of recruiting a limited fishing ground in process of exhaustion, by planting artificially hatched larvæ on it. The most economical method of obtaining the eggs would be from fish caught in the course of commercial fishing. Obviously the scale on which the hatchery would most effectively work would be determined by a knowledge of the annual reduction of the fish population by fishing operations.

Explanation of Plates.

Reference Letters.

A. car.—Internal carotid arteries (the two outer ones).

A. car'.—External carotid artery.

A. Cm.—Coeliaco—mesenteric artery.

A. coe.—Coeliac artery.

A. ep.—Right epibranchial artery.

A.F.—Anal fin.

A*f. Br. 1—4.*—First to fourth afferent branchial arteries.

A. gen.—Common genital artery.

A. gen'—Right genital artery.

A. hep.—Hepatic artery.

A. hy.—Hyoidean artery.

Al. S.—Alisphenoid.

amp. a.
amp. e. ⎫—Anterior, external, and posterior ampullæ of the semicircular
amp. p. ⎭   canals.

An.—Angular.

a. nos.
a. nos'. ⎬—Right and left anterior nostrils.

ant. cr.—Crista acustica ampullæ anterioris.

Ao. *d.*—Dorsal aorta.

A. œ.—Oesophageal artery.

A. op.—Ophthalmic artery.

Ao. *V.*—Ventral aorta.

A. pe.—Pelvic artery.

Ar.—Articular.

A.R.—Accessory ribs or intermuscular bones, numbered from before backwards.

A.R.*—Tubercle on fourteenth and many of the succeeding vertebræ, representing probably a fused accessory rib.

a. s. c.—Anterior semicircular canal.

A. scl.—Right and left subclavian arteries.

A. sp.—Splenic artery.

Aur.—Auricle.

Ax.—Axonosts or interspinous bones, numbered from before backwards.

Ax.ª—Transverse ligament connecting axonost with skin.

Ax.ᵇ—Triangular plug of cartilage occurring in the head of every axonost.

A. Zy.—Anterior zygapophysis.

B. *A.*—Bulbus arteriosus.

B. *Br. 1-4.*—Basi-branchials 1 to 4.

Bd.—Bile duct.

B. *O.*—Basi-occipital.

Br. 1.—First holobranch. *1*, the first visceral arch.

Br. R.—Branchiostegal rays, numbered from before backwards.

Bs.—Baseosts, numbered from before backwards.

C.—Centrum.

C. Br.^{1}
C. Br.^{4} }—Cerato-branchials of the first and fourth branchial arches.

C. F.—Facet on atlas for paroccipital condyle.

C. Fn.—Caudal fin.

C. Hy.—Cerato-hyal.

cil. b.—Ramus ciliaris brevis.

cil. g.—Ciliary ganglion.

cil. l.—Ramus ciliaris longus.

Cir. c.—Circulus cephalicus.

Cl.—Clavicle.

Co.—Coracoid.

com. 1—vii..—RR. communicantes between the sympathetic and the first seven spinal nerves.

com. v.^{1}—R. communicans between the sympathetic and the N. trigeminus.

Cr. C.—Cranial cavity.

d. 2—6.—Dorsal roots of the spinal nerves 2 to 6.

D.—Dentary.

d. b.
d. c. }—Dorsal roots of the first spinal nerve.

d. end.—Vestigial ductus endolymphaticus, with two minute utriculo-saccular canals opening into its base.

D. F.—Dorsal fin.

E. Br.^{1}
E. Br.^{4} }—Epibranchials of the first and fourth branchial arches.

Ef. Br. 1—4.—First to fourth efferent branchial arteries.

Ep. 1
Ep. 2 }—First and second epural bones.

Ep. Hy.—Epi-hyals.

Ep. O.—Epiotic.

Ep. P.—Epiotic process of epiotic.

e. s. c.—External (horizontal) semicircular canal.

Eth.—Ethmoid cartilage.

Ex. O.—Exoccipital.

ext. cr.—Crista acustica ampullæ externæ.

f. car.—Carotid foramen, transmitting the internal carotid artery.

f. gl.—Foramen of the N. glossopharyngeus.

f. jug.—Jugular foramen, transmitting the superior (internal) jugular vein, the ophthalmic artery and the T. hyomandibularis facialis.

F. M.—Foramen magnum (occipital foramen).

f. olf.—Olfactory foramen in the pre-frontal by which the N. olfactorius reaches the nasal chamber.

F.R.—Fin rays.

F.R.$^{a.b}$ —The two pieces forming a single fin ray.

F.R.^{c} —Ligament connecting the two halves of the fin ray.

F.R.^{d} —Ligament connecting one half of the fin ray with the baseost.

F.R.^{e} —The right abductor muscle of the fin ray. The elevator and depressor muscles do not appear in this section.

F. Sp. N.—Foramina for the spinal nerves.

F. Sp. N.^{3} —Foramina for the fourth spinal nerve.

f. tr. fa.—Trigemino-facial foramen, transmitting the vth and viith nerves except the T. hyomandibularis facialis.

f. vg.—Foramen of the N. vagus.

g. 3—6.—Ganglia of the spinal nerves 3 to 6.

g. v—vii.—Massed ganglia of the trigeminal and facial nerves.

g. ix.—Ganglion of the N. glossopharyngeus.

g. x. 1.—Ganglion of the first branchial trunk of the N. vagus.

g. x. 2—5.—Ganglionic complex formed by the second, third and fourth branchial + the intestinal ganglia of the N. vagus.

g. coel.—Ganglion coeliacum.

g. d. 2.—Dorsal ganglion of the second spinal nerve.

g. extcr.—Extracranial ganglion of the first spinal nerve.

g. intcr.—Intracranial ganglion of the first spinal nerve.

g. v. 2.—Ventral ganglion of the second spinal nerve.

H. A.—Haemal arches, numbered from before backwards.

H. Br.^{1} —Hypo-branchial of the first branchial arch.

H. C.—Haemal canal.

H. Hy.—Hypo-byals.

Hm.—Hyomandibular.

Hm. F.$^{1-2}$—Cup and facet for head of hyomandibular. Cp. fig. 5.

Hp.[1]
Hp.[2] }—First, second and third hypural bones.
Hp.[3]

H. *S.*—Haemal spines, numbered from before backwards.

H. *S.*†—Last three haemal spines.

hy. c.—Hyomandibular sensory canal.

I. Cl.—"Inter-clavicle."

I. Hy.—Inter-hyal.

I. M. C.—Inter-maxillary cartilage.

In.—Innominate or pubic bone. The figures denote its three parts.

in. buc.—R. buccalis internus facialis.

inf. c.—Infraorbital sensory canal.

I. Op.—Inter-operculum.

I. Ph.—Inferior pharyngeal bones, representing a fifth pair of branchial arches.

Jac. anast.—Jacobson's anastomosis from the ninth to the seventh cranial nerves.

Jug. g.—Jugular ganglion of the N. vagus.

lag.—Lagena.

lat. c.—Lateral sensory canal.

L. Fr.—Left frontal.

l. g. x.—Ganglion of the R. lateralis vagi.

L. Lc.—Left lachrymal.

L. Na. C.—Position of left nasal chamber.

low. in. buc.—Lower division of the R. buccalis internus facialis.

L. P. C.—Left paroccipital condyle.

L. P. Fr.—Left pre-frontal.

l. rec. vii.
l. rec. x. }—R. lateralis recurrens facialis and vagi.

man. v.
man. vii. }—R. mandibularis trigemini and facialis.

man. ext. vii.—R. mandibularis externus facialis.

man. int. vii.—R. mandibularis internus facialis.

m. a. s.—Macula acustica sacculi.

M. *C.*—Pads of soft cartilage (fibrous in parts).

m. d. op.—Twig from the N. trigeminus to the M. depressor operculi.

M. *E.*—Posterior portion of the mesethmoid bounding the orbit in front.

M. *E.'*—Anterior portion of the same bearing the beak-like ridge for the intermaxillary cartilage.

M. *Pt.*—Meta-pterygoid.

m. r. u.—Macula acustica recessus utriculi.

Ms. Pt.—Meso-pterygoid.

Mx.—Maxilla.

mx. v.—R. maxillaris trigemini.

N.—Cup shaped cavity at each end of centrum containing vestiges of the notochord.

N. A.—Neural arch.

Nc.—Notochord.

N. C.—Notochordal canal in centrum placing the notochordal spaces (*N.*) in communication.

n. ch.
n. ch.^{1} }—Right and left nasal chambers.

n. olf.
n. olf.^{1} }—Right and left olfactory nerves.

N. S.—Neural spines, numbered from before backwards.

N. S.†—Last three neural spines.

n. sac.$^{1-5}$—The five right and left nasal sacs.

n. sp.'
n. sp.'' }—*NN.* splanchnici.

O.C.—Occipital condyle.

Oc. S.—Occipital spine.

Od.—Opening of oviduct.

Od.'—Common ovarial chamber.

Od.''—Interior of right ovary.

o. i.—Branch of the N. oculomotorius to the inferior oblique muscle.

o. lam.
o. lam.' }—Right and left series of olfactory laminæ.

Op.—Operculum.

Op. O.—Opisthotic.

o. pr.—Nervus ophthalmicus profundus (trigemini ?) and ganglion.

op. s. vii.—R. opercularis superficialis facialis.

o. s.—N. patheticus to superior oblique muscle.

oto.—Saccular otolith.

out. buc.—R. buccalis externus facialis.

Pa.—Palatine.

pal.—R. palatinus facialis.

p. a. l.—Papilla acustica lagenæ.

Pa. P.—Parotic process of pterotic and opisthotic.

Par.—Parietal.

Pa. S.—Parasphenoid.

P. *Br.*[1] —Pharyngo-branchial of the first branchial arch articulating with the skull.

P. *Br.*[2-4]—Pharyngo-branchials of the branchial arches two to four, forming the superior pharyngeal bone (*S. Ph.*).

P. *C.*—Paroccipital condyle.

P. *Cl.*—Post-clavicle.

Per.—Pericardium.

P. *F.*—Pectoral fin.

ph. x. 1
ph. x. 2 }—RR. pharyngei of the first, second and third branchial trunks of the vagus.
ph. x. 3

Pl. F.—Pelvic fin.

Pl. F.'—Base of right pelvic fin.

P. *Mx.*—Premaxilla.

p. nos.
p. nos.[1] }—Right and left posterior nostrils.

P. *Op.*—Pre-operculum.

post. 1
post. 2 }—RR. post-trematici of the first, second, third, and fourth branchial
post. 3   trunks of the vagus.
post. 4

post. vii.—R. post-trematicus facialis.

post. ix.—R. post-trematicus glossopharyngei.

post. cr.—Crista acustica ampullæ posterioris.

pre. 1.
pre. 2. }—RR. pre-trematici of the first, second, third and fourth
pre. 3.   branchial trunks of the vagus.
pre. 4.

Pr. O.—Prootic.

Ps. Br.—Pseudobranch.

p. s. c.—Posterior semicircular canal.

Pt.—Pterygoid.

Pt. O.—Pterotic.

P. *Tp.*—Post-temporal.

P. *Zy.*—Posterior zygapophysis.

Qu.—Quadrate.

r. v.—Root of N. trigeminus.

r. 1. vii.—Dorsal (sensory) roots of N. facialis.

r. 2. vii.—Ventral (motor) root of N. facialis.

r. ix.—Root of N. glossopharyngeus.

r. x.—Root of N. vagus *sensu stricto.*

R. —Ribs, numbered from before backwards.

r. a. a.—R. acusticus ampullæ anterioris.

r. a. e.—R. acusticus ampullæ externæ.

r. a. p.—R. acusticus ampullæ posterioris.

r. car. ?—R. cardiacus vagi ?

r. cerv.—R. cervicalis of the first spinal nerve ($=$ " N. hypoglossus.")

r. com. 2—6.—RR. communicantes of spinal nerves 2 to 6.

r. com. b.—R. communicans of the first spinal nerve.

r. e.—N. abducens to external rectus.

Rec. Ax.[1] —Longitudinal furrow on the anterior aspect of the first haemal spine for the reception of the first axonost of the anal fin.

R. Fr.—Right frontal.

r. hy.—R. hyoideus facialis.

r. if.—Branch of N. oculomotorius to rectus inferior.

r. intest. x.—R. intestinalis vagi.

r. it.—Branch of the N. oculomotorius to rectus internus.

r. l.—R. acusticus lagenæ.

r. lat. x.—R. lateralis " vagi."

r. lat. x.'—Root of R. lateralis vagi.

r. lat. prof. x.—R. lateralis profundus vagi.

r. lat. sup. x.—R. lateralis superficialis vagi.

R. Lc.—Right lachrymal.

r. m. 2—5.—RR. medii of spinal nerves 2 to 5.

r. m. b.
r. m. c. } —RR. medii of the first spinal nerve.

R. Na.—Right nasal.

R. Na. C.—Position of right nasal chamber.

r. op. x.—R. opercularis vagi.

r. oph. sup. v.—R. ophthalmicus superficialis trigemini.

r. oph. sup. vii.—R. ophthalmicus superficialis facialis.

r. ot.—R. oticus facialis.

R. P. Fr.—Right prefrontal.

r. r. u.—R. acusticus recessus utriculi.

r. s.—Branch of N. oculomotorius to rectus superior.

r. sac.—R. acusticus sacculi.

r. sp. 2—5.—RR. spinosi of spinal nerves 2 to 5.

r. sp. b.
r. sp. c. } —RR. spinosi of the first spinal nerve.

r. st. x.—R. supratemporalis vagi.

r. v. 1.—R. ventralis of the first spinal nerve + an anastomosis from the second.

$\left.\begin{array}{l} r.\ v.\ 1.^1 \\ r.\ v.\ 1.^2 \end{array}\right\}$ Motor branches of the brachial plexus.

$\left.\begin{array}{l} r.\ v.\ 1.^3\text{—Motor} \\ r.\ v.\ 1.^4\text{—Sensory} \end{array}\right|$ branches of the brachial plexus to the pectoral fin.

r. v. 2.'—Anastomosis between RR. ventrales of the first two spinal nerves.

r. v. 2 + 3.—Fused RR. ventrales of spinal nerves 2 and 3.

r. v. 2 + 3.'—Motor branch sending an anastomosis to R. ventralis 4.

r. v. 2 + 3″.—Sensory nerve to pectoral fin.

r. v. 2—5.—RR. ventrales of spinal nerves 2 to 5.

r. v. b + c.—R. ventralis of first spinal nerve.

rx. b.—Radix brevis.

rx. l.—Radix longa ganglii ciliaris (accompanied by sympathetic).

sac.—Sacculus.

Sc.—Scapula.

S. C.—Spinal canal.

S. Cl.—Supra-clavicle.

Sin. V.—Sinus venosus.

Sk.—Outline of skull.

S. O.—Supra-occipital.

s. o. c.—Supra-orbital commissure between the two supra-orbital sensory canals.

S. Op.—Sub-operculum.

S. Ph.—Superior pharyngeal bone of the eyeless side intact. The numbers denote the branchial arches to which its constituents belong.

Spl.—Spleen.

Sp. N.[13]—Thirteenth spinal nerve.

Sp. O.—Sphenotic.

S. R.—Strengthening ridges of the vertebral centra.

s. t. c.—Supratemporal portion of the lateral sensory canal.

sup. c.—Right supraorbital sensory canal.

$\left.\begin{array}{l} sup.\ c.' \\ sup.\ c.'' \end{array}\right\}$—Vestiges of the left supraorbital sensory canal.

Sy.—Symplectic.

$\left.\begin{array}{l} t.\quad .\ 1. \\ t.\quad .\ 2. \\ t.\ \overset{x}{z}.\ 3. \end{array}\right\}$—First, second and third Trunci branchiales vagi.

R

242

Tb. 1—5.—The five tuberosities seen externally on the ocular side passing
backwards from between the two eyes.

Thm.—Thymus gland.

t. hm.—Truncus hyomandibularis facialis.

t. inf.—Truncus infraorbitalis (trigemini et facialis).

Tr. P.—Transverse process.

U. + Hp. 3.—Urostyle and third hypural fused together.

U. Hy.—Basi-hyal.

up. in. buc.—Upper division of the R. buccalis internus facialis.

Ur.—Urostyle.

Uret.—Ureter.

Ur. pp.—Urinary papilla.

utr.
utr.' }—Central and vertical chambers of the utriculus.

v. 2—6.—Ventral roots of spinal nerves 2 to 6.

V.—Vertebral centra numbered from before backwards. V^1 is the atlas.

v. b.'
v. b. }—Ventral roots of the first spinal nerve.
v. c.

V. card.—Cardinal vein.

Ven.—Ventricle.

V. gen.—Genital vein.

V. hep.—Hepatic vein.

V. jug.
V. jug.' }—Superior and inferior jugular veins.

Vo.—Vomer.

Vp.—Portal veins.

V. pc.—Right precaval vein.

Plate I.

Fig. 1. Dorsal surface of an undried skull of a 65cm.
Plaice. Natural size. The dotted line indicates the departure from the symmetry. The
anterior extremity is somewhat schematic,
and is drawn as if seen a little from behind.
The fenestra in the ethmoid cartilage is only
partially visible in a strictly dorsal view.

Fig. 2. Ventral surface of the same undried skull. Natural size. The dotted line indicates the swerve of the ventral axis of the cranium towards the left side caused by the rotation towards the right of the orbital region of the cranium. The chondrocranium appears on the surface of the cranium in several places, and not always symmetrically.

Fig. 3. Lateral view of the same skull seen from the right or ocular side. Natural size. The right nasal and right lachrymal are here fully shown, and do not appear in perspective as in figs. 1 and 2.

PLATE II.

Fig. 4. The occipital region of the same skull viewed from behind. Natural size. The dotted line indicates the departure from the symmetry. The chondrocranium appears on the surface between the exoccipitals and epiotics.

Fig. 5. Lateral (moist) dissection of the opercular bones, palato-pterygoid arcade, and jaw apparatus of the Plaice, on the ocular or right side. Natural size. Extreme length of specimen 34cm. The palatine has been rotated downwards in order to illustrate its shape, and the lower jaw has been depressed. In the natural disposition of the parts the lower end of the maxilla lies external to the mandible.

Fig. 6. Dorsal (moist) dissection of the Hyoid arch of a 34cm. Plaice. Natural size. In the living animal the stout hyoid bar is almost vertical and the basi-hyal projects forwards at right-angles from its upper border. The lower border of the bar in the figure is therefore the dorsal border, and the face shown is the anterior face. The first pair of branchiostegal rays has been turned forwards. They really pass straight backwards in the mid-ventral line. The branchiostegal rays are numbered from before backwards, and have been displayed.

Fig. 7. Dorsal (moist) dissection of the branchial arches of a 34cm. Plaice. Natural size. The arches have been flattened out and displayed. On the right side the three pharyngo-branchials forming the superior pharyngeal bone have been separated and left attached to their respective arches, but on the left that bone is represented intact drawn from its oral surface. The arches are numbered from before backwards (i.-v.).

Fig. 8. Lateral (moist) dissection of the right pectoral girdle and fin of a 34cm. Plaice. Natural size. The " inter-clavicle " in front is shown in its correct position relative to that of the pectoral girdle. The fin rays have been separated from the two fibro-cartilaginous pads with which they articulate.

Fig. 9. Lateral (moist) dissection of the right pelvic girdle and fin of a 34cm. Plaice. Natural size. The fin rays are represented disarticulated from the fibro-cartilaginous pad.

Plate III.

Fig. 10. First or atlas vertebra viewed from the front of a 52cm. Plaice. Natural size. The accessory ribs (intermuscular bones) have been rotated forwards so that their exact length may be indicated. The dotted line shows the extent of the asymmetry.

Fig. 11. The eighth trunk vertebra viewed from the front of a 64cm. Plaice. Natural size. The accessory ribs have been rotated forwards in order to introduce their full length. The dotted line shows the departure from the symmetry.

Fig. 12. The twelfth trunk vertebra of a 52cm. Plaice seen in optical longitudinal section, the ocular or right half of the centrum and right neural arch having been ground away. Natural size. The figure illustrates the amphicoelous character of the centrum, and the shape and extent of the notochordal spaces.

Fig. 13. The fourteenth or first caudal vertebra of a 52cm. Plaice seen from the front. Natural size. The dotted line indicates the departure from the symmetry. The same vertebra is shown in side view in fig. 18.

Fig. 14. Moist dissection from right side of caudal fin and termination of vertebral column of a 34cm. Plaice. Natural size. The fin rays have been disarticulated from the fibrocartilaginous pad.

Fig. 15. Dissection from right side of posterior extremity of notochord and caudal fin supports of a young Plaice of 17mm. Drawn with camera. x 15. Notochord as yet unossified, but faint vertebral constrictions (6 shown in fig.) have appeared. The neural and haemal spines and epural and hypural bones are beginning to ossify from without inwards. The fig. should be compared with fig. 14. It shows that the urostyle has not yet fused with the third hypural, and that the caudal fin of the young Plaice is of the ordinary homocercal type.

Fig. 16. Longitudinal section through a fin ray and axonost of the dorsal fin of a medium sized adult Plaice, *i.e.*, the section passing longitudinally through the fin ray and transversely to the long axis of the fish. Leitz 3, ocular 2. The figure shows the two pieces forming one fin ray, and their relations and connections with the baseost and axonost. The section does not pass through the plane of the elevator and depressor muscles of the fin ray, or the longitudinal vertical ligament of the axonosts.

PLATE IV.

Fig. 17. Moist dissection from the eyeless or left side of the anterior part of the dorsal fin and first three trunk vertebræ of a 52cm. Plaice. Natural size. The outline of the skull is introduced to show its ventral inclination and its relation to the first eight-ten axonosts of the dorsal fin. (Cp. table in text.) The first three vertebræ also have a downward inclination.

Fig. 18. Moist lateral dissection from the eyeless side of the last four trunk and first three caudal vertebræ, and of the anterior extremity of the anal fin and its supports of a 52cm. Plaice.

Two-thirds natural size. The ring **at** the ventral extremity of axonost **1** indicates how much of the latter protrudes freely from the body wall in dead specimens. Cp. table in text.

Fig. 19. Moist lateral dissection from the eyeless side of the last seven caudal vertebræ and the posterior extremities of the dorsal and anal fins and their supports of a 52cm. Plaice. Natural size. The sigmoid curve of fin ray **71** of the dorsal fin and its horizontal position are abnormalities. Cp. with fig. **14** and table in text.

PLATE V.

Fig. 20. The viscera seen from the ocular side. Only the body wall has been removed and the muscles of the limb girdles dissected away. The ovary and rectum are opened to shew the external opening of the oviduct. The specimen was a ripe female Two-thirds natural size.

Fig. 21. The viscera seen from the ocular side. The body wall and the superficial coils of the intestine have been removed and the pectoral girdle, operculum, sub-operculum and inter-operculum dissected away. The lateral wall of the pericardium has been removed. The positions of haemal spine 1 and axonost 1 are indicated by broken lines. The specimen was a spent female. Two-thirds natural size.

Plate VI.

Fig. 22. Diagrammatic representation of the principal blood vessels and their distribution. The principal viscera are introduced into the figure in order to shew the distribution of the vessels only. Their space relations to each other are not accurately shewn. The intestine is represented for simplicity's sake as a single coil. The hepatic portal system is not shewn, nor the distribution of the carotid arteries. The arteries are cross hatched, the veins are uniformly shaded. About natural size.

Plate VII.

Fig. 23. Reconstruction from serial sections of the cranial nerves of the ocular side. The two scales in this and other figures refer to the numbers of the sections. Sensory canals coloured. green. Their sense organs and surface pores are numbered in each canal from before backwards. Eye muscle nerves cross striated. The numerals refer to the numbers of the cranial nerves. Note the relatively large size of the eyes in the young fish. The curve of the medulla shown in this figure does not exist in the adult, and is probably an abnormality.

Plate VIII.

Fig. 24. Reconstruction of the left ear and auditory nerve seen from the inner or left side. This figure cannot be compared with fig. 23, the reconstruction being made on a somewhat larger scale, although from the same series of sections. The nerve ramuli in front are a little diagrammatic, in order that they may be shown more clearly. viii., auditory nerve. The reconstruction applies also to the adult ear, but the relative sizes of the parts are of course different.

Fig. 25. Reconstruction of the right and left nasal organs seen from *above*. Drawn to exactly twice the scale of figs. 23 and 26, and from the same series of sections. The disposition of the accessory secretory chambers is slightly diagrammatic for the sake of clearness. The termination of the olfactory nerves is not shown for the same reason.

Fig. 26. The sympathetic nervous system seen from above, and plotted from the same series of sections and to the same scale as fig. 23. The dotted line represents the median line of the body continued straight forwards. The figure is diagrammatic in four respects: (1) the cœliac ganglion lies directly under ganglion 2; (2) the first cranial ganglion lies immediately over the second on the ocular side; (3) the ocular ciliary ganglion lies over the third nerve; (4) the eyeless ciliary ganglion is external to the oculomotorius. The cranial ganglia are numbered 1-8, the spinal 1'-7'. iii. is the Nervus oculomotorius, after it has given off the nerve to the M. rectus superior.

PLATE IX.

Fig. 27. Reconstruction from serial sections of the first

s

six spinal nerves of the right side seen from the lateral aspect. The two crosses indicate the position of the foramen magnum. The cranial region is to the right of these, the vertebral to the left. The reconstruction is from the same series of sections as figures 23-26, but is not drawn to the same scale. The formal shading of the ganglia is omitted in the case of the first spinal for the sake of clearness. The roots of most of the spinal nerves lie internal to the ganglia and therefore cannot be shown. Their places of origin, however, are indicated by the oval dotted areas.

Fig. 28. Transverse section through the fore-brain to illustrate the asymmetry of this part of the brain. The dotted line indicates the mid vertical plane. Most of the brain therefore here lies to the right. Note the fibres collecting at the base of each bulbus olfactorius. These ultimately form the olfactory nerves. Drawn with the camera, Zeiss aa. oc.4.

PLATE X.

Fig. 29. Plaice split up the middle from below into two halves and spread out to show both sides of the sensory canal system. x$\frac{3}{4}$. The anterior extremity of the dorsal fin is not shown. The shading indicates the innervation of the sensory canals. *Supraorbital canals*, cross hatched (R. ophthalmicus superficialis, vii.); *Infraorbital canals*, dotted (R. buccalis, vii.); *Hyomandibular canals*, shaded (R. mandibularis externus, vii.); *Lateral canals*, cross striated (R. lateralis, **x.**). The line of dashes represents the suppressed portion of the left supraorbital canal.

Fig. 30. Dorsal view of the brain of a Plaice having an extreme length of 58cm. x 3. The pallium and the choroid roof of the fourth ventricle have been removed, but the asymmetrical position of the pineal tube and body is indicated. The dotted line shows the departure from the symmetry. The numbers are those of the cranial nerves.

Fig. 31. Lateral view of the same brain with the pallium *in situ.* x 3. Numbers as in preceding figure.

PLATE XI.

Fig. 32. Embryo Plaice on the 5th day after fertilization and before invagination of the optic vesicles. × 29.

Fig. 33. Embryo Plaice on the 9th day after fertiliza tion. The black spots on the body and tail represent the yellow branching chromato phores which appear about this time. × 29.

Fig. 34. Embryo Plaice on the 17th day after fertilization. The embryo is ready to hatch. The branching black markings on the body and tail are the black chromatophores. The yellow pigment is represented by the grey spots. × 29. Nat. size of this and the preceding eggs = 1·88 mm.

Fig. 35. Newly hatched Plaice, 17 days after fertilization. The black pigment is represented bv the stellar markings, the yellow bv the grey spots. × 17. Nat. length = 6·3 mm.

Figs. 32—35 are drawn from the living eggs

and larva from material treated at the Piel Hatchery. The average temperature of the water in the hatching boxes during the developmental period was 7° C.

Fig. 36. Larval Plaice at the beginning of the metamorphosis. The yolk-sac has been absorbed, and the larva is beginning to feed. Note the beginning of the asymmetry—the left eye is more dorsal than the right. Note also the upturned end of the notochord. Nat. size = 15 mm. × 6. (After Ehrenbaum, Eier und Larven von Fischen der Deutschen Bucht. Wiss. Meeresuntersuch. Kiel. Komm. Bd. 2, Heft I., Abth. I., Plate 4, fig. 15. 1896.)

Fig. 37. First bottom stage of the Plaice. The metamorphosis is completed. Note the relatively large size of the eyes and the small paired fins. × $4\frac{1}{4}$. Nat. length = 13 mm. (After Petersen, 4th Rep. Danish Biological Station. Pl. I., fig. 10. 1893.)

Tinling & Co., Printers, 53, Victoria Street Liverpool.

Vo
R Na C.
R.Na.
R Lc
R.P.Fr.
M.E.
L.Lc.
L.Fr.
Tb. 1.
Ocular Side
Tb. 2.
R.Fr.
Sp.O.
Tb 3
Par.
S.O.
Par.
Pt.O.
Tb. 4.
Ep.O.
Ep.O.
Ex O

Fig.

R Lc.
R Na.C.
R.Na.
Vo.
L Na C
L Lc
f.olf.
R.P.Fr.
L.P.Fr.
Eth.
Pa.S.
Ocular Side
R.Fr.
Eyeless Si
L.Fr.
L.Fr.
Cr.C.
Pa.S.
Al.S.
Al.S.
f.tr.fa.
f.tr.fa.
Sp.O.
Sp.O.
f.jug.
f.car.
Hm.F.
f.car.
f.jug.
Pr.O.
Pa.S.
Pr.O.
Hm.F.
Pt.O.
B.O.
B.O.
f.gl.
f.gl.
Op.O.
Op.O.
Pt.O.
Ex.O.
B.O.
Ex.O.
Ex O
f.vg.
O.C.
P.C.
f.vg.

Fig. 2.

Par.
R.Fr.
L Fr
R Na C
M.E.
R.Na.
M.E.
L PFr
Sp.O.
Sp.O.
R.Fr.
R.P.Fr.
R Lc.
Hm.F.
R.Fr.
f.olf.
L.Fr.
L.P.Fr.
Al.S.
Pa.S.
Vo.
f.jug.
f.tr.fa.
Pr.O.
Pa.S.
Eth.
L.P.Fr.
f.car.
Pa.S.
Ocular Side

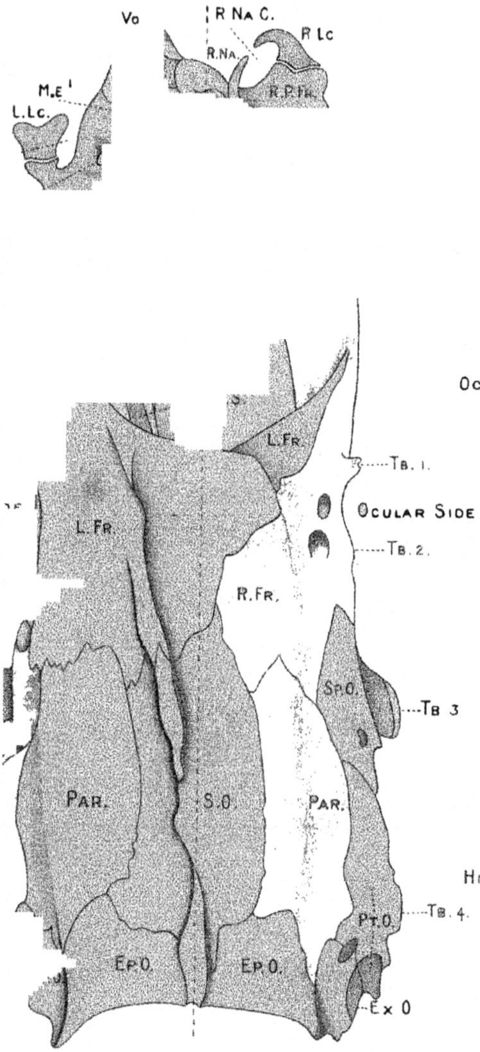

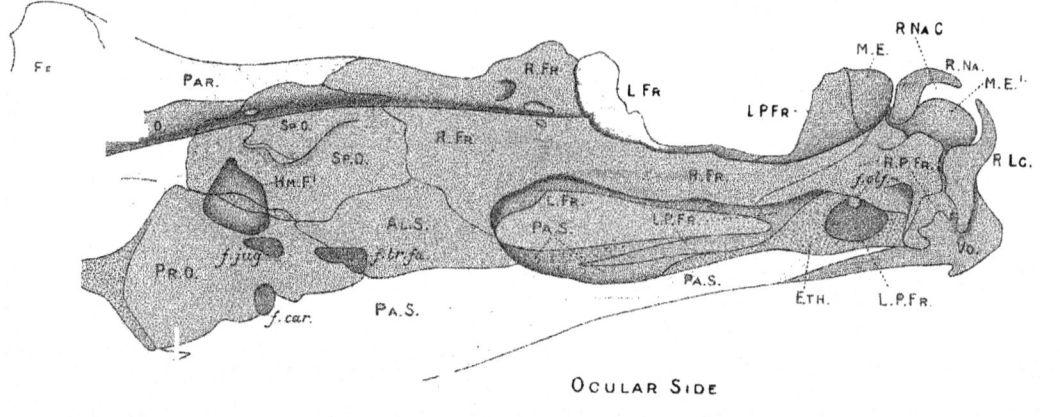

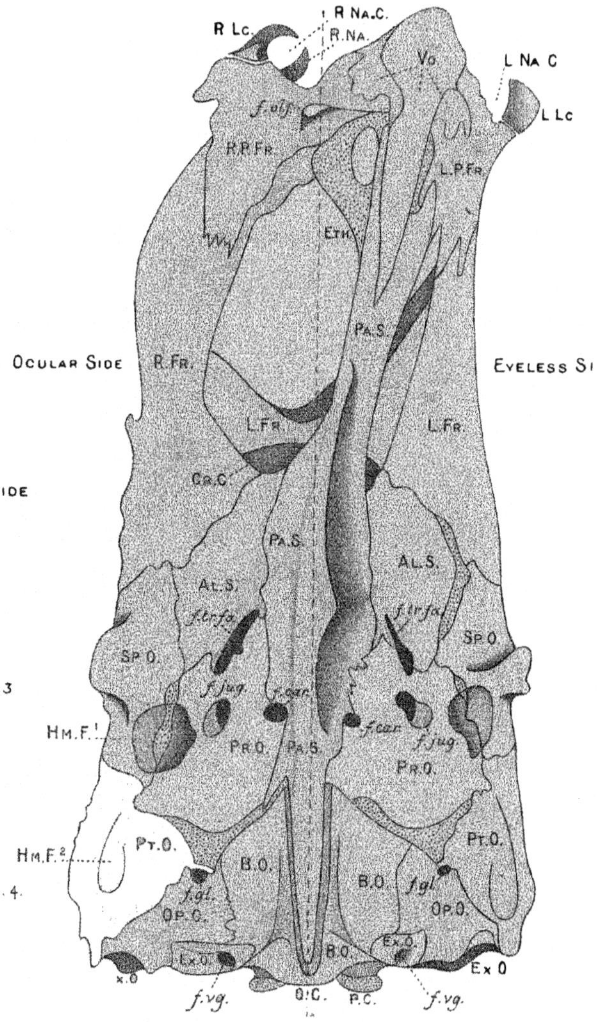

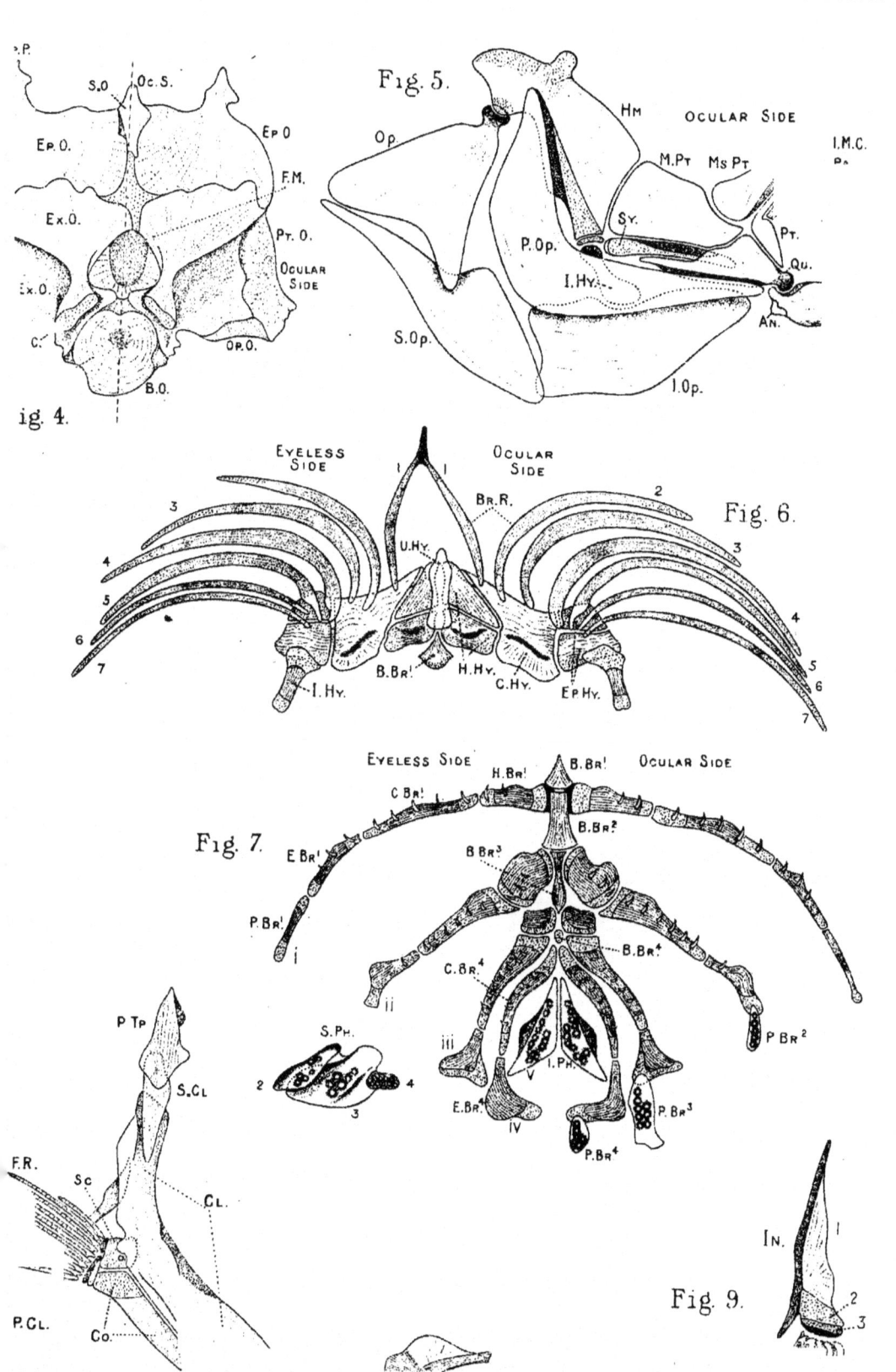

Fig. 4.

Fig. 5.

OCULAR SIDE

Fig. 6.

EYELESS SIDE

OCULAR SIDE

Fig. 7.

EYELESS SIDE

OCULAR SIDE

Fig. 9.

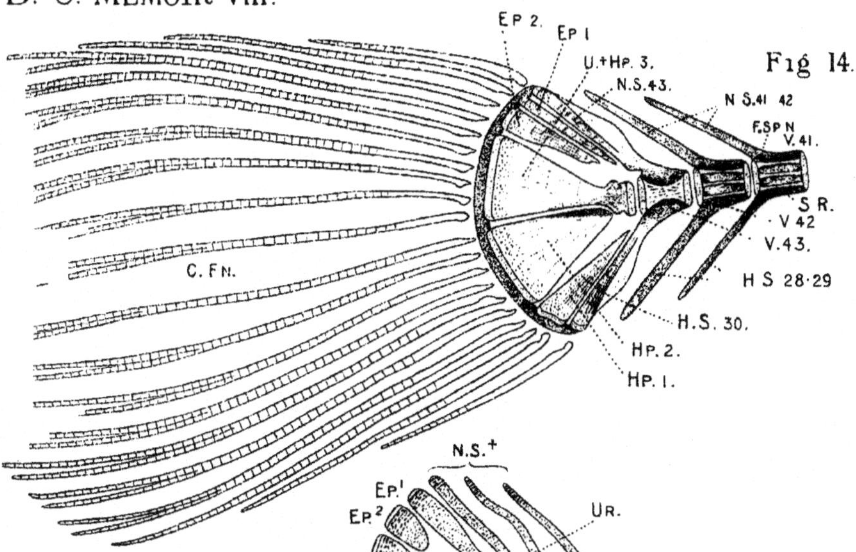

Fig 14.

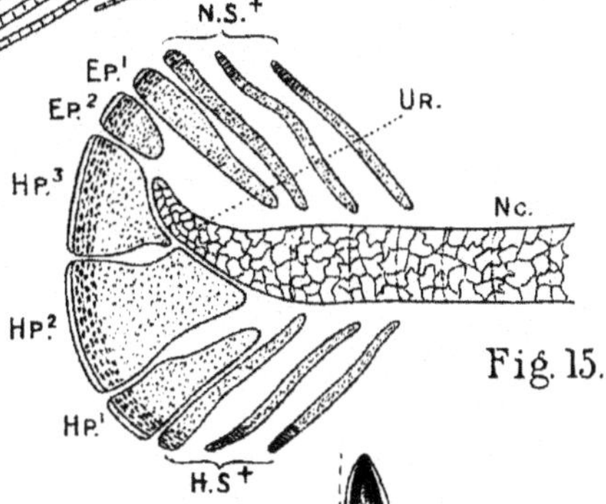

Fig. 15.

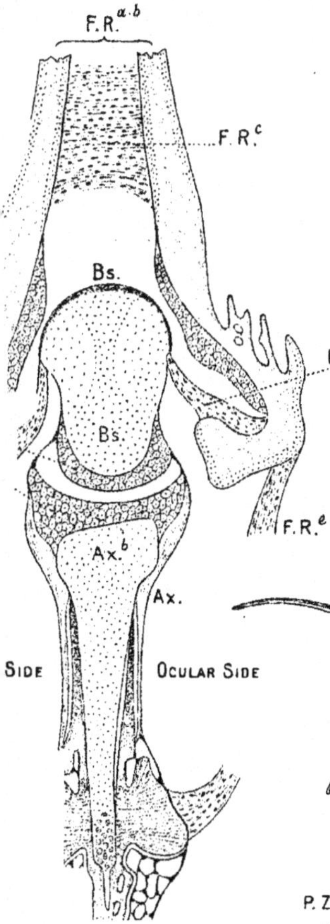

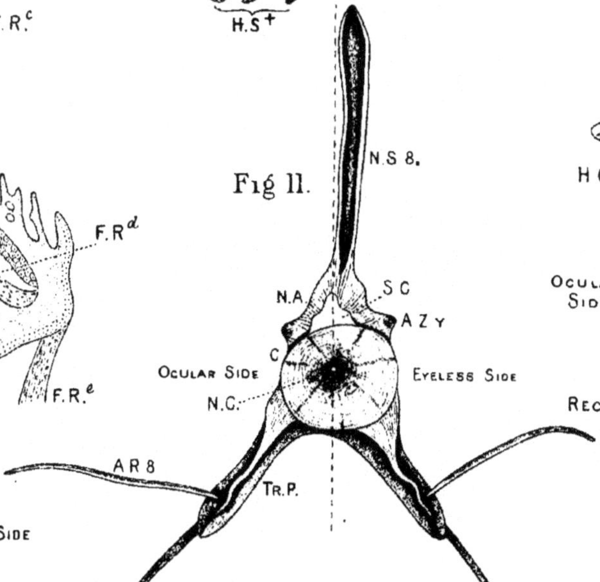

Fig 11.

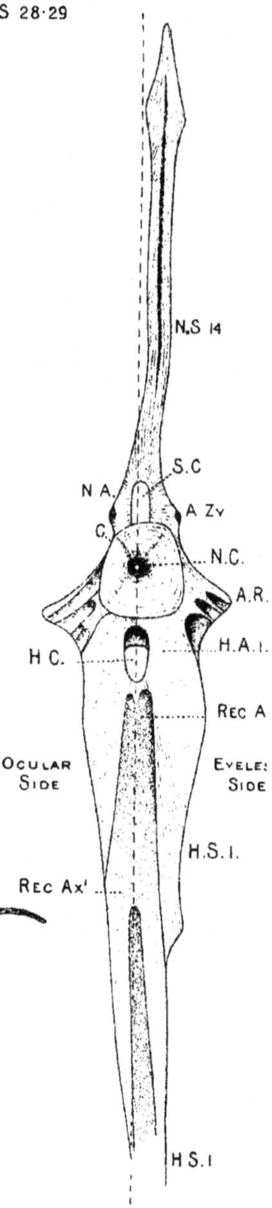

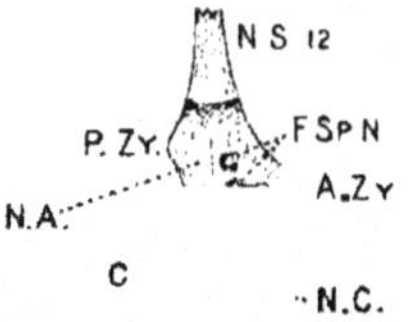

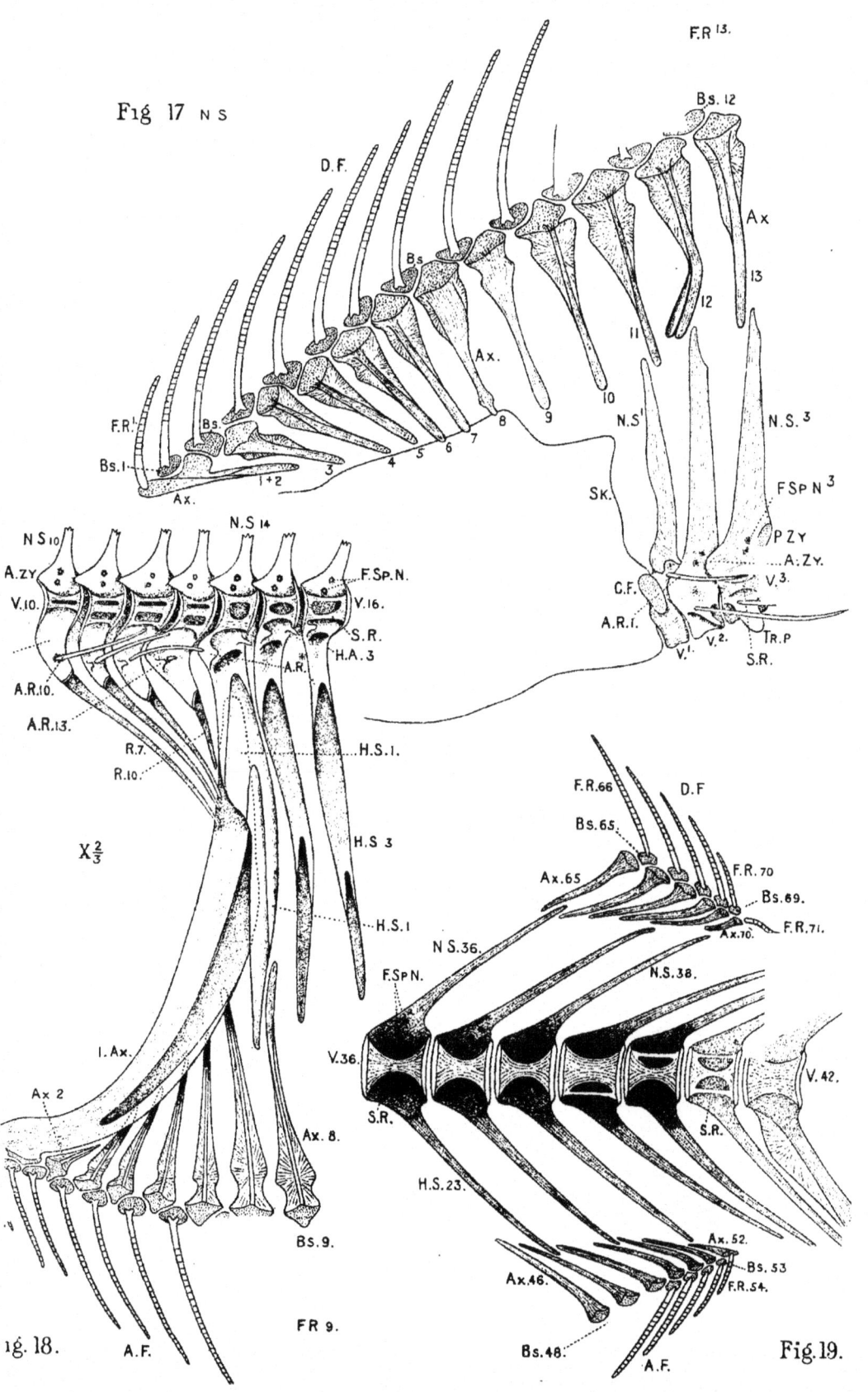

Fig 17 NS

F.R 13.

D.F.

Bs. 12

Ax

13

Bs

Ax.

12

F.R¹

Bs.

11

Bs.1

10

N.S'

N.S.³

1+2

Ax.

3

4 5 6 7 8 9

Sk.

FSPN³

PZY

A.ZY

V.3.

NS 10

N.S 14

C.F.

A.ZY

F.Sp.N.

V.10.

V.16.

A.R.1.

V.1 V.2.

TR.P

S.R.

S.R.

H.A.3

A.R.10.

A.R.

A.R.13.

H.S.1.

R.7.

H.S 3

R.10.

F.R.66

D.F

Bs.65.

F.R.70

X⅔

H.S.1

Ax.65

Bs.69.

Ax.70.

F.R.71.

N.S.36.

F.Sp.N.

N.S.38.

1.Ax.

V.36.

V.42.

Ax 2

S.R.

S.R.

Ax. 8.

H.S.23.

Ax.52.

Bs.53

Bs. 9.

Ax.46.

F.R.54.

ig. 18.

A.F.

FR 9

Bs.48.

A.F.

Fig.19.

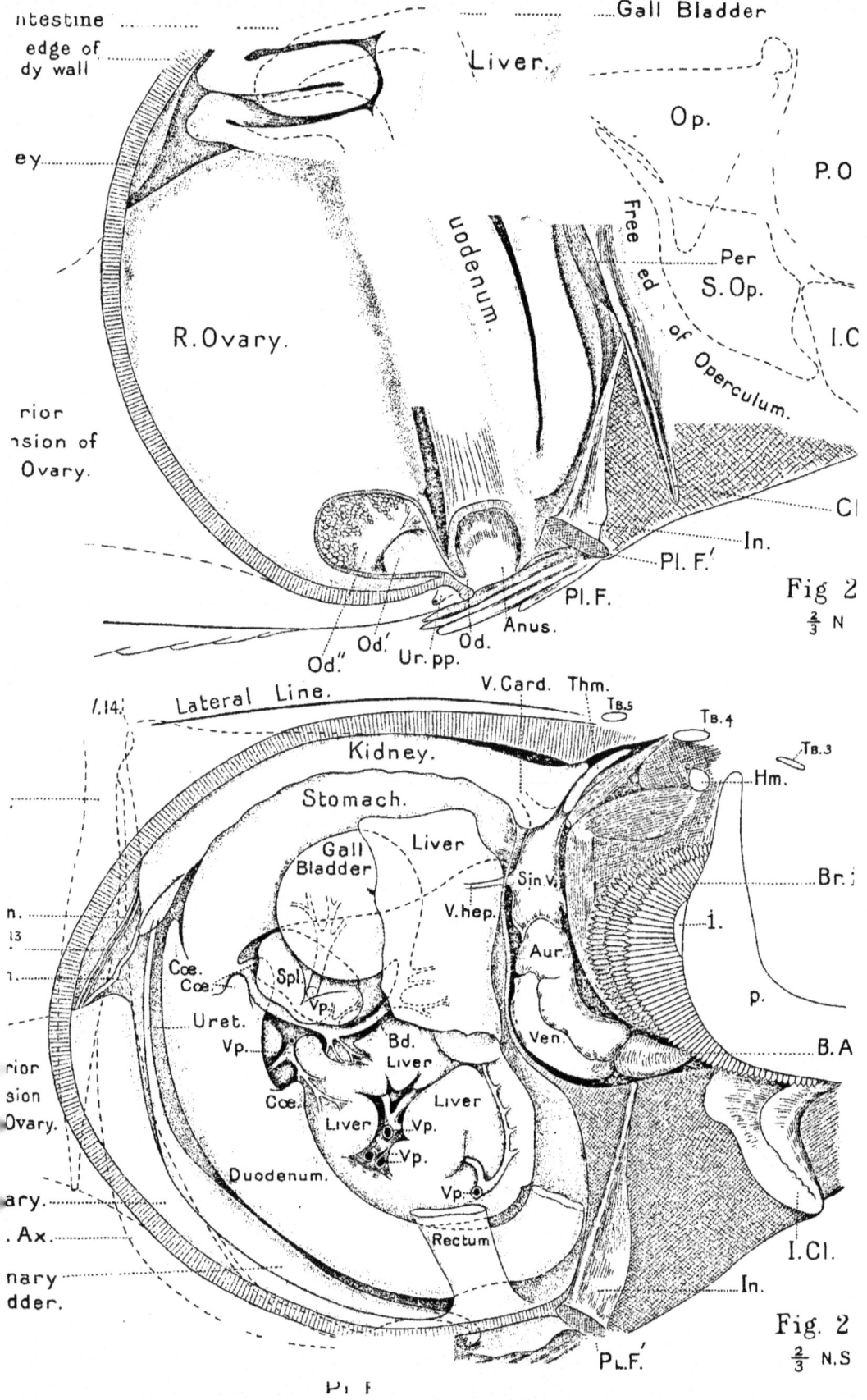

ntestine
edge of
dy wall

ey

rior
nsion of
Ovary.

Gall Bladder

Liver.

Op.

P.O

Per

S.Op.

I.C

of Operculum.

Free ed

uodenum.

R.Ovary.

Cl

In.

Pl. F.'

Pl. F.

Anus.

Od." Ur. pp. Od. Od.'

Fig 2
$\frac{2}{3}$ N

Lateral Line.

V. Card. Thm.

Tb.5

Tb.4

Tb.3

l.14.

Kidney.

Hm.

Stomach.

Gall
Bladder

Liver

Sin.V.

Br.

V.hep.

n.

l3

Coe.
Coe.

Spl

Aur.

l.

Vp.

Uret.
Vp.

Bd.
Liver

Ven.

p.

B.A

Coe.

Liver

Liver

Vp.

rior
sion
Ovary.

Liver

Vp.

Duodenum.

Vp.

ary.

.Ax.

Rectum

I.Cl.

nary
dder.

In.

Fig 2
$\frac{2}{3}$ N.S

P.L.F.'

Pl F

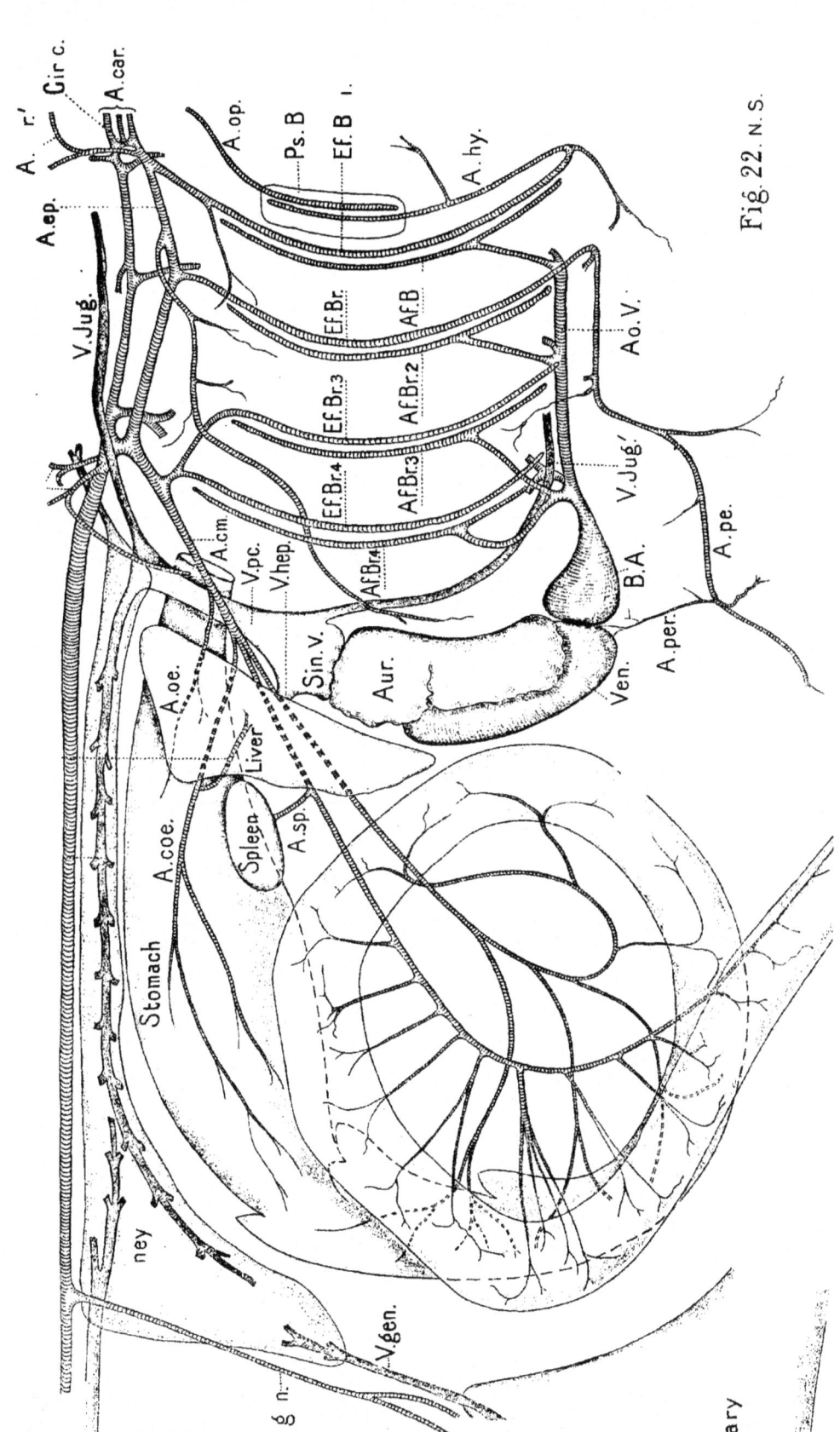

A. r' Cir c. A.car.
A. A.sp.
A.op.
Ps. B
Ef. B l.
A. hy.
V.Jug.
Ef.Br. AFB
Ef.Br.3 AfBr2
Ef.Br.4 AfBr.3 Ao.V.
A.cm AfBr.4
V.pc. V.hep.
Af.Br.4 V.Jug.'
Sin. v. B.A. A.pe.
Aur. A.per.
A.oe. Ven.
Liver
Spleen
A.sp.
A.coe.
Stomach
ney
V.gen.
g n.
ary

Fig 22 N.S.

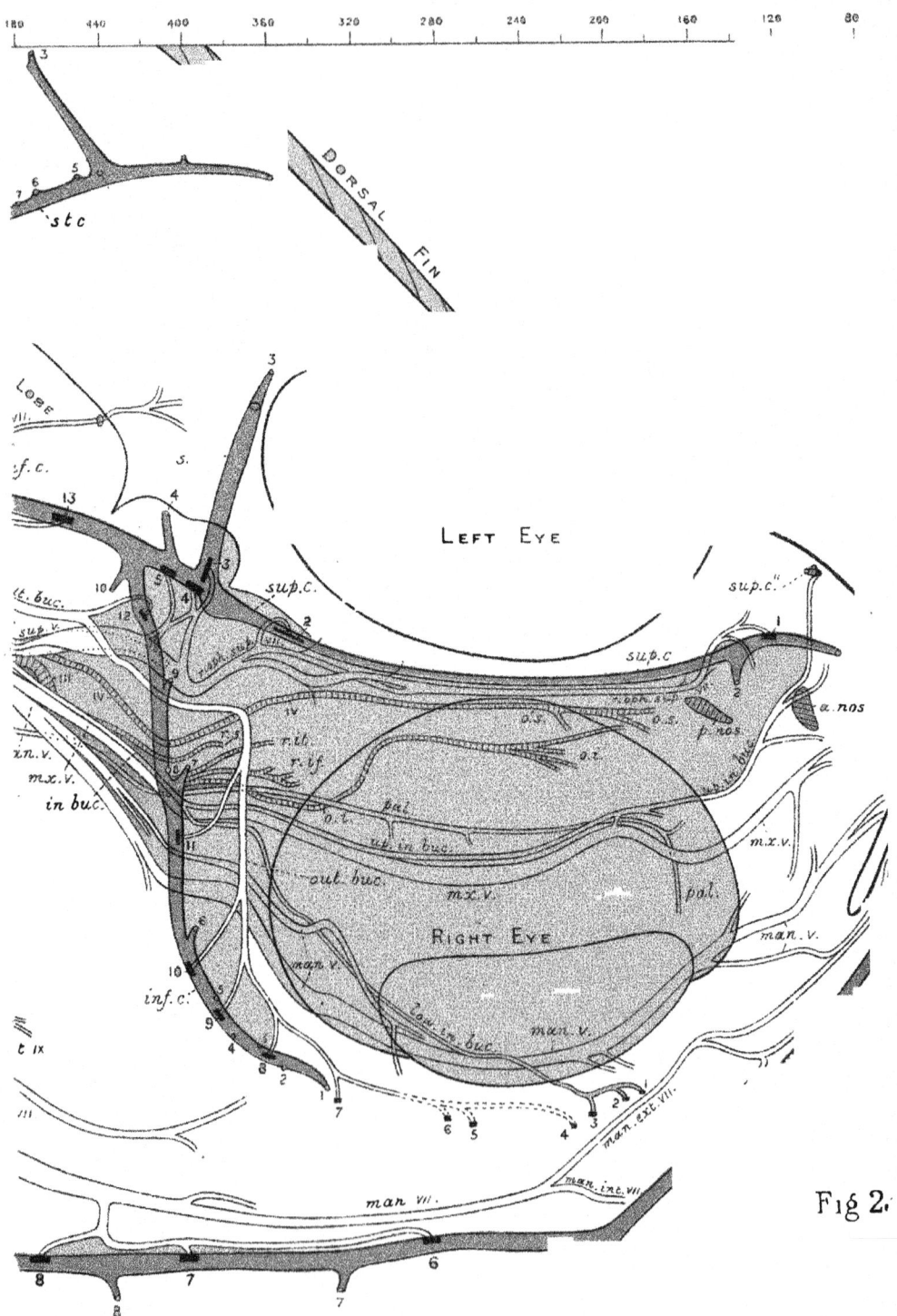

Fig 2.

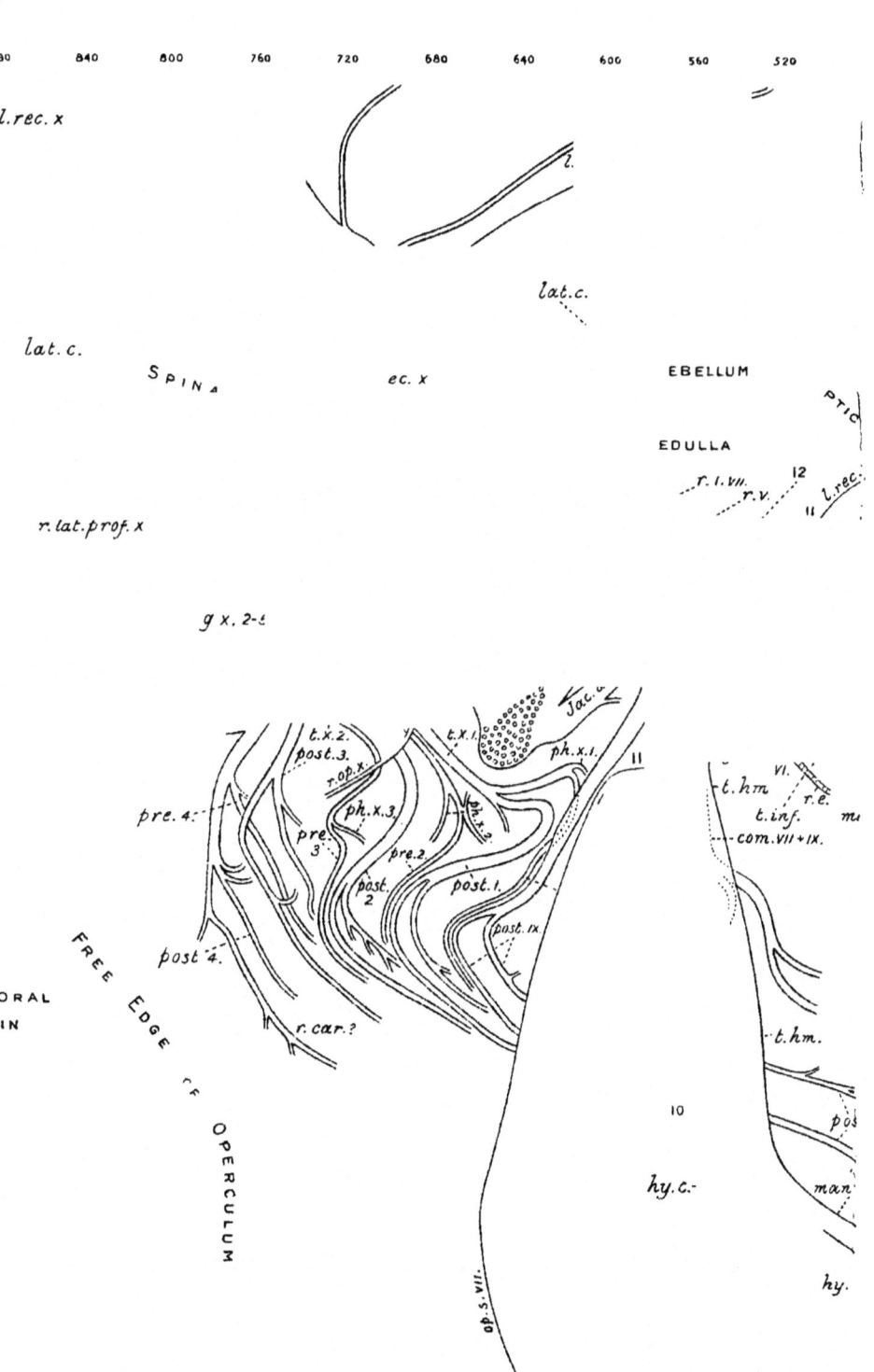

880 840 800 760 720 680 640 600 560 520

l.rec. x

l.

lat.c.

lat.c.

S P I N A   *ec. x*   EBELLUM   PTIC

EDULLA

r. l. vii.   12

r. v.   *l. rec.*

11

r. lat. prof. x

g x. 2-

t.x.2.   *t.x.1.*   Jac.

post.3.   *ph.x.1.*   11

r.op.x.   *t.hm*   VI.

pre.4.   *ph.x.3.*   *ph.x.*   *r.e.*

pre.   *t.inf.*   m

3   *pre.2.*   com.vii + ix.

post.   *post.1.*

2

post. ix

post 4.

FREE   *r. car.?*

TORAL   EDGE   *t.hm.*

FIN   O F

10   *pos*

O P E R C U L U M   *man*

hy.c.

op. s. vii.

hy.

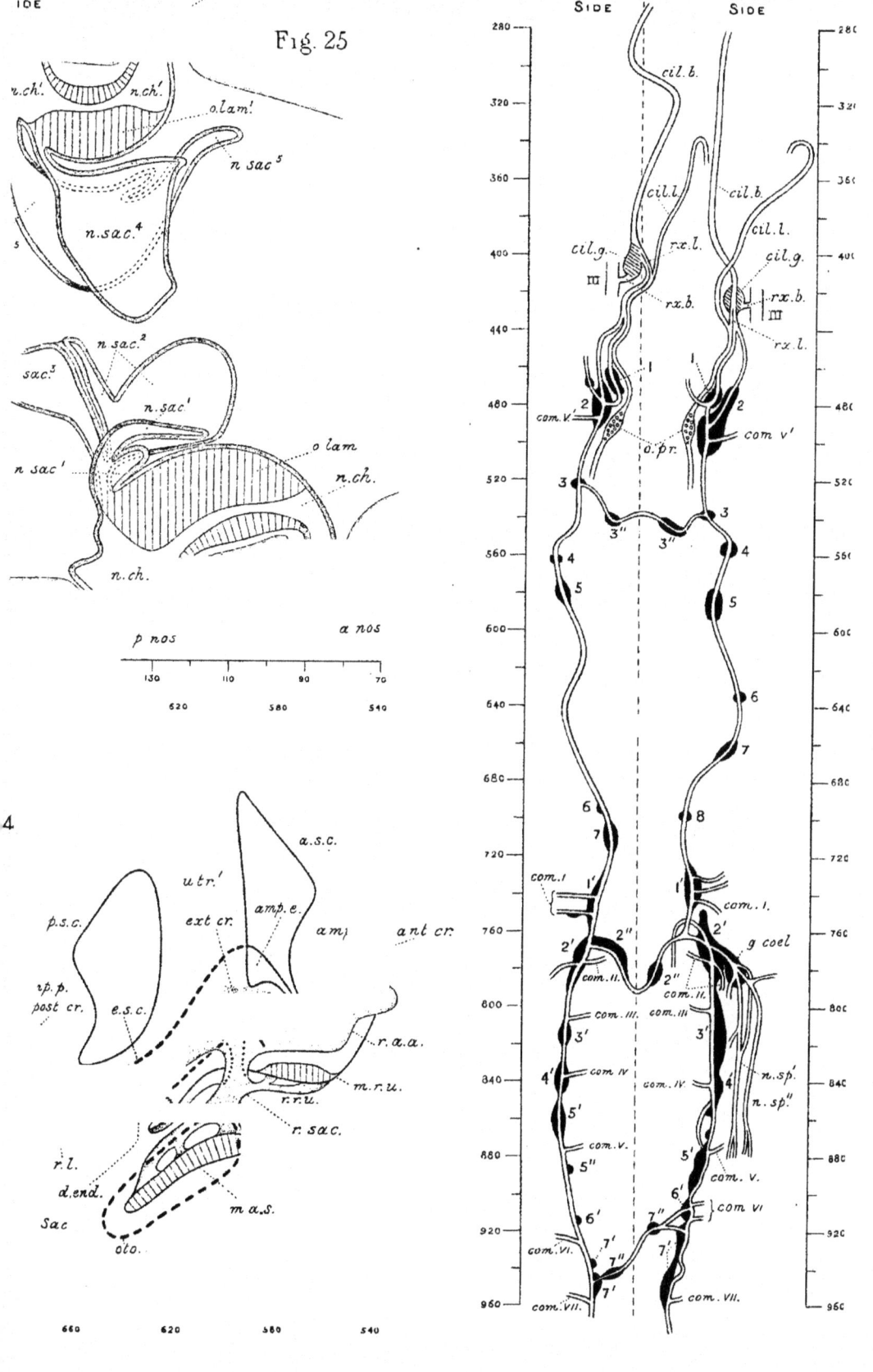

Fig. 25

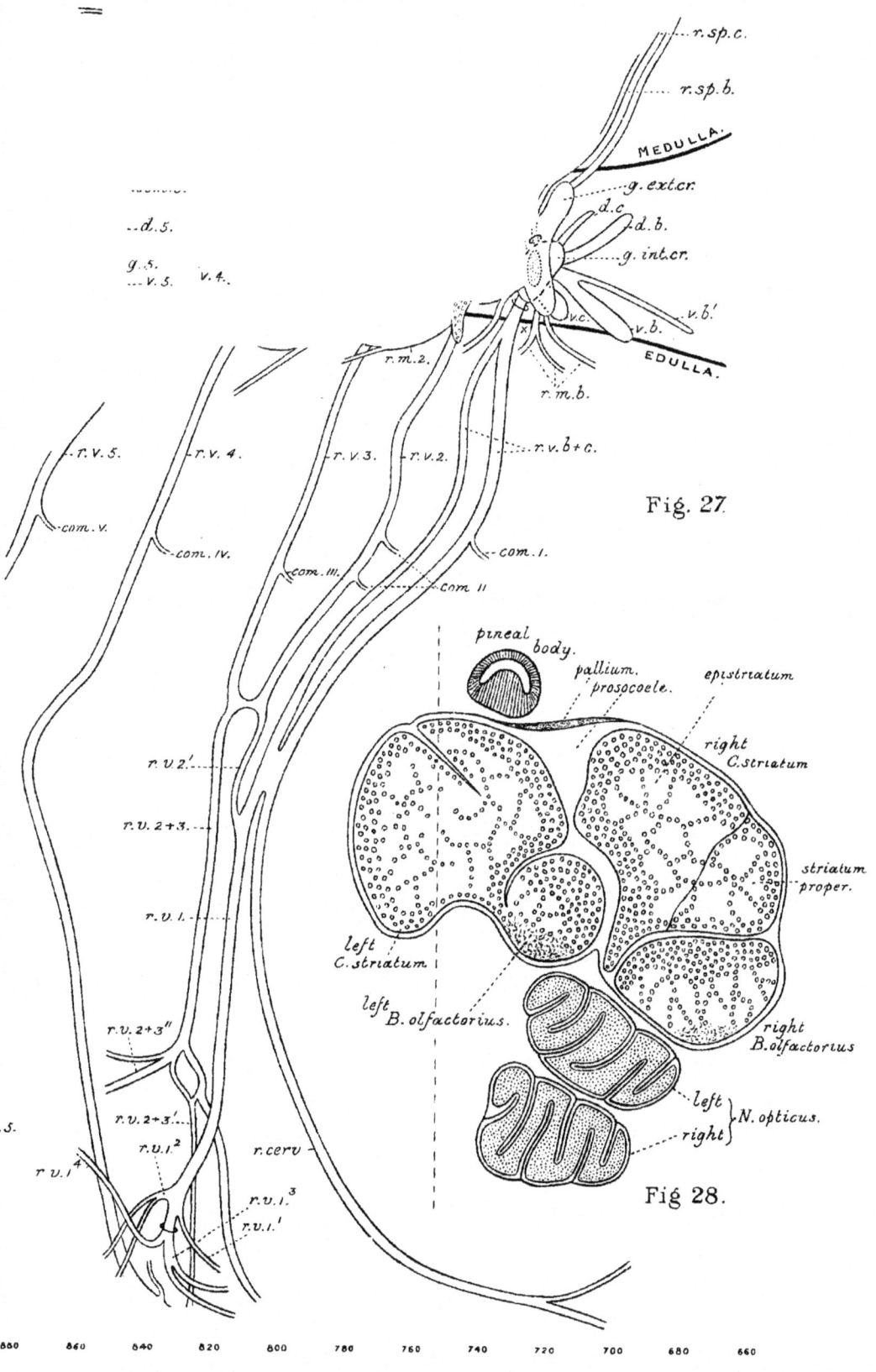

r.sp.c.

r.sp.b.

MEDULLA.

g.ext.cr.

d.c

d.b.

g.int.cr.

v.c.

v.b.'

v.b.

EDULLA.

r.m.2.

r.m.b.

...d.5.

g.5.

...v.5.   v.4.

r.v.5.   r.v.4.   r.v.3.   r.v.2.   r.v.b+c.

com.v.   com.IV.   com.III.   com.II   com.I.

Fig. 27.

r.v 2'.

r.v.2+3.

r.v.1.

r.v.2+3"

r.v.2+3'.

.s.

r.v.1²   r.cerv

r.v.1⁴   r.v.1³

r.v.1¹

pineal body.

pallium.

prosocoele.

epistriatum

right C.striatum

striatum proper.

left C.striatum

left B.olfactorius.

right B.olfactorius

left } N.opticus.

right }

Fig 28.

880   860   840   820   800   780   760   740   720   700   680   660

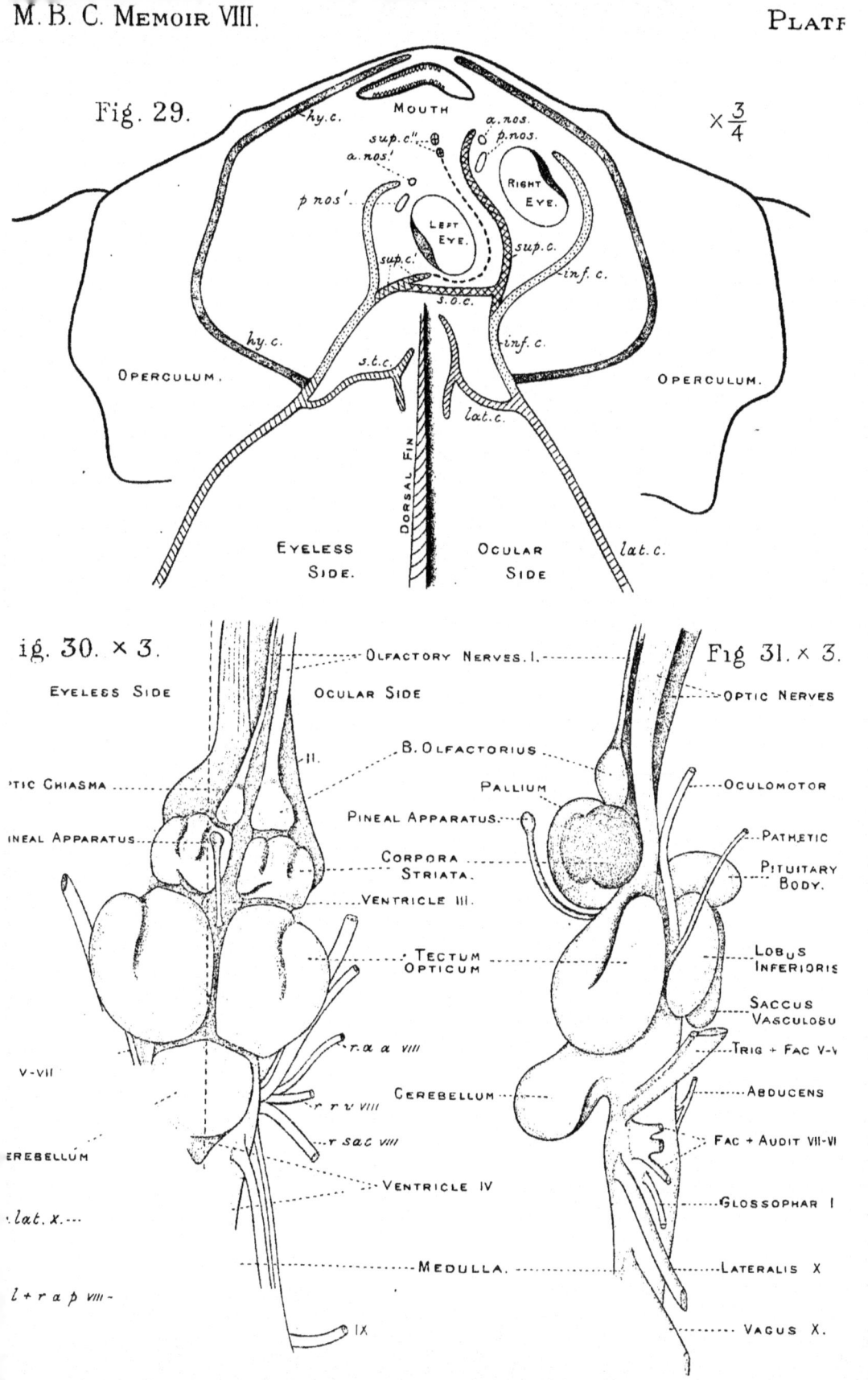

Fig. 29.

× ¾

MOUTH

hy.c. a.nos. p.nos.

sup.c''. a.nos'.

p.nos'. RIGHT EYE.

LEFT EYE. sup.c.

sup.c. inf.c.

s.o.c.

hy.c. inf.c.

s.t.c. lat.c.

OPERCULUM. OPERCULUM.

lat.c.

EYELESS SIDE. OCULAR SIDE

DORSAL FIN

Fig. 30. × 3. Fig 31. × 3.

EYELESS SIDE OCULAR SIDE

OLFACTORY NERVES. I. OPTIC NERVES

II. B. OLFACTORIUS

OPTIC CHIASMA PALLIUM OCULOMOTOR

PINEAL APPARATUS. PATHETIC

PINEAL APPARATUS. PITUITARY BODY.

CORPORA STRIATA.

VENTRICLE III.

TECTUM OPTICUM LOBUS INFERIORIS

SACCUS VASCULOSU

TRIG + FAC V-V

r.a.a VIII

V-VII CEREBELLUM ABDUCENS

r.r.v VIII

r.sac VIII FAC + AUDIT VII-VI

CEREBELLUM

VENTRICLE IV GLOSSOPHAR I

lat.x.

l + r.a.p VIII MEDULLA. LATERALIS X

IX VAGUS X.

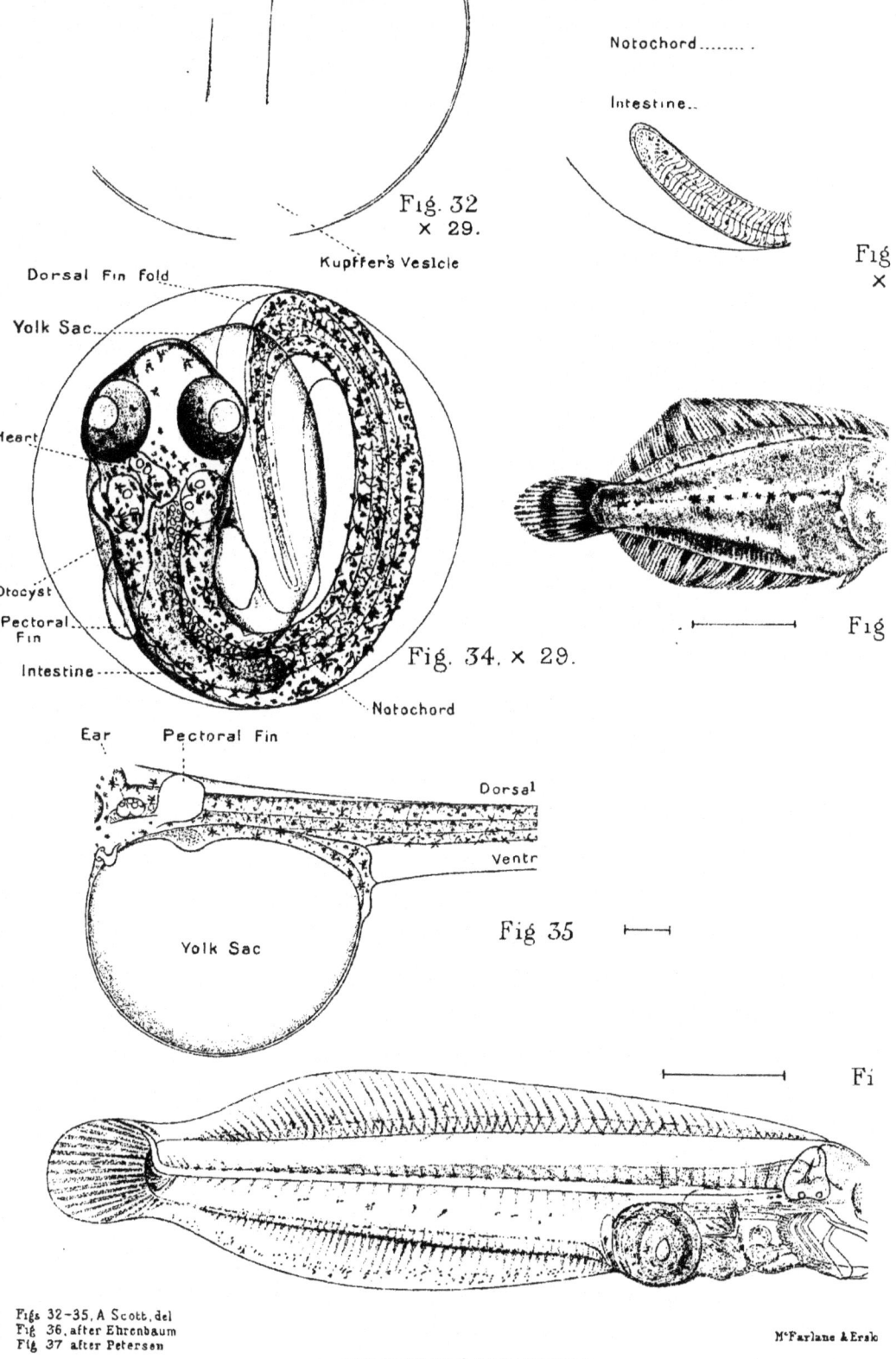

Notochord........ .

Intestine..

Fig. 32
× 29.

Kupffer's Vesicle

Fig
×

Dorsal Fin fold

Yolk Sac...

Heart

Otocyst

Pectoral Fin

Intestine....

Fig. 34. × 29.

Notochord

Fig

Ear   Pectoral Fin

Dorsal

Ventr

Fig 35

Yolk Sac

Fi

PLEURONECTES

Lightning Source UK Ltd.
Milton Keynes UK
UKOW05f0606050617
302702UK00016B/835/P